Ordinal Data

Ordinal Data

NONPARAMETRIC STATISTICAL ANALYSES AND
SPSS APPLICATIONS

CLIFTON W. HAMILTON

BARRY UNIVERSITY

SAN DIEGO

Bassim Hamadeh, CEO and Publisher
Christina Brown, Associate Acquisitions Editor
Skyler Van Valkenburgh, Project Editor
Casey Hands, Production Editor
Emely Villavicencio, Senior Graphic Designer
Laura Duncan, Licensing Coordinator
Natalie Piccotti, Director of Marketing
Kassie Graves, Senior Vice President, Editorial

320 South Cedros Ave., Ste. 400, Solana Beach, CA 92075

Table of Contents

Preface

It is my hope that this handbook helps students in the field of health and wellness and the behavioral sciences to recognize and understand the properties of ordinal data. Based on my teaching experience, many students in these fields will be engaged in projects that include the use of Likert-like data, rankings, ordinal measures, ordered preferences, and nonparametric data sets, so I tried to include as many of these types of data in examples and practice exercises as space would allow. I also tried to use simple, introductory language throughout the handbook; at the same time, it is expected that readers have some prior knowledge of basic statistical concepts such as the null hypothesis, type I and type II errors, p values, random sampling, and populations. Since these and other statistical concepts often need reviewed, I included a list of key terms and their definitions at the beginning of each unit when appropriate.

In some of the results sections, I use the phrase "we have evidence that ..." or "we lack evidence that ..." to remind students that the results of one statistical analysis alone do not establish a truth or fact. Results of analyses, for the most part, either support or do not support research hypotheses. Results of studies are like grains of sand that accumulate into a pile on the beach, representing the accumulation of knowledge from which we draw our conclusions about a topic or phenomenon. These conclusions are about certain populations that should be defined at the beginning of any research endeavor. While we often use a random sample from the population because of convenience, results are in reference to the population from which the sample was randomly chosen.

The data sets used for instructional purposes and as practice exercises are, for the most part, unusually small. This was done to save space and to lessen the burden of entering large amounts of data into a hand calculation or software program. Some data sets are fictitious while others are excerpts of actual data sets used for various studies. I tried to make the fictitious data sets as realistic as possible. All described procedures for conducting the various tests in SPSS are based on version 29. A realistic timetable for reading this textbook is six to seven weeks.

Finally, I want to acknowledge those individuals who gave me valuable advice on various concepts and procedures. Rick Marcantonio and Jon Peck from IBM's Quality Assurance team were able to clarify for me some output terminology used in SPSS, and Doug Wolfe, professor emeritus (Ohio State University), explained in clear terms the limitations of nonparametric testing. My professors at Florida International University, Dr. Leonard Bliss and Dr. Abbas Tashakkori, were instrumental in introducing me to and helping me begin to understand inferential statistics.

Clifton W. Hamilton
Los Angeles County, California

Introduction to Ordinal Data

OBJECTIVES

- Understand the characteristics of ordinal data.

- Understand the usefulness of ordinal data.

- Analyze tools that collect ordinal data.

- Understand the importance of a tool's validity and reliability.

KEY TERMS

Ordinal data: Data in which the observations are ordered or ranked, without standardized differences between order or rank. Two examples of ordinal data are high school class ranking and Likert scaling scores.

Discrete observations: Observations of categories or rankings that are exact, whole values, not made into fractions or into negative values. Both nominal data (categorical) and ordinal data (ranks) are discrete.

Interval data: Data in which the observations are numbers, with the distance between numbers assumed to be equal but discrete (whole numbers) with no true zero. Examples include temperature in Celsius or Fahrenheit (something is occurring at zero) and some standardized tests, when the test constructor posits that each item is worth equal value.

Scale (ratio) data: Data that meet the criteria of interval data but also have a true zero.

Validity: The ability of a tool to measure what it purports to measure.

Reliability: A measure of consistency regarding subjects' scores across a construct.

A BRIEF INTRODUCTION TO THIS UNIT

It would be difficult to pinpoint the exact date when humans first began to record ordinal data. By the sixth century BCE, chariot races were taking place in the Circus Maximus, and we know that Roman officials kept track of both the winner and subsequent placing of chariots, their drivers, and their horses. Turning to more modern times for the use of ordinal data, the results of the third race run at Belmont Park on September 29, 2022 (*Daily Racing Form*) show Jackson Heights winning by two lengths and Arctic Annoyance placing second over Donegal Surges by 3¾ lengths. But for the bettors of that race, what was most important is who won, who placed (came in second), and who showed (came in third). That Arctic Annoyance lost by two lengths might be important for a bettor interested in Arctic Annoyance's next race, but as we use ordinal data for research purposes, we will not assume that distances between ordinal measures are either calculable or of interest.

With ordinal data, we cannot always establish standardized differences between measures or ranks, but more on that later. Because future bettors of horse races are also concerned about how much a horse won by or lost by, the *Daily Racing Form* also includes those lengths of separation in its results. We will not delve into such details in this handbook. Another common use of ordinal data appears in the form of anchors for Likert-like items. In a recently released film, a prison psychologist asks an inmate to respond to a question with, to the best of my memory, either *strongly disagree, disagree, unsure, agree,* or *strongly agree*. It was an intense scene, and the inmate in question had been incarcerated for 12 years for having committed a violent crime on a family member. While I was slightly amused by the use of Likert scaling in popular culture, I am not so sure that it is fitting to ask an inmate with a violent past to respond to Likert-like items about his own violent acts. My point here is that, while ordinal data can be very useful, the mind-set of the respondent and their willingness to answer honestly is crucial to drawing valid and reliable conclusions about the topic under study. This point will be elaborated on in the following sections.

THE USEFULNESS OF ORDINAL DATA

Tools that collect ordinal data allow us to measure phenomena of the behavioral sciences in quantitative terms, often in a quick, valid, and reliable manner. Ordinal data help to describe levels of comfort, agreement, aggression, or authoritarianism; opinions on topics such as the effectiveness of a program, a medicine, a training regimen, or a doctor; and knowledge or awareness of pollution, crime, police activity, or governmental policy. For decades, the US Census Bureau, the US Postal Service, and practically every department of education across the nation has collected ordinal data to measure levels of satisfaction and concerns on a variety of topics for such efforts as redesigning programs and policymaking. Agricultural education is a prolific user of ordinal data in measuring levels of satisfaction and agreement.

When it comes to quantifying sensory events or emotions in the behavioral sciences with ordinal data, credit must be given to Rensis Likert, whose publication in 1932 set the standard for the development of the eponymous Likert scale. A Likert-like scale consists of a series of statements or questions (items) all related to one construct (e.g., trust in my neighbor). The quantification of the construct should be the sum of *all* scores on each item found in the survey or tool. With that said, it is argued here that all items should be worded in such a way so that the summative score is reflective of trust in one's neighbor. To clarify, let us examine a fictitious item (table 1.1).

TABLE 1.1 **Fictitious Likert-like Item**

I trust my neighbor will feed my cats when I am out of town if I ask them to.

1 = strongly disagree	2 = disagree	3 = neutral	4 = agree	5 = strongly agree

If the researcher wishes to measure one's trust in their neighbor, then all items on the survey should be constructed as in table 1.1. That is, a score of 5 should indicate the highest level of trust on any item. If this "trust my neighbor" survey contained ten items similar in design, the highest summative score a respondent could achieve would be 50 and the lowest summative score a respondent could achieve would be 10. According to van Sonderen et al. (2013), it would be a mistake to include a negatively worded item in the same survey as table 1.2.

TABLE 1.2 **Questionable Fictitious Likert-like Item for Measuring One's Trust in Their Neighbor**

I do not trust my neighbor to water my roses when I am out of town if I ask them to.

1 = strongly disagree	2 = disagree	3 = neutral	4 = agree	5 = strongly agree

When a reverse-oriented item is used in a survey, it might be there because the survey designer wished to prevent *response bias*—the tendency to respond to items regardless of the item's content. Ostensibly, this ploy forces the respondent to become more engaged with the survey. In such a case, the researcher would also have to reverse the scoring to maintain an honest summative score—in the case of the item in table 1.2, awarding the respondent a value of 1 if they chose 5 (strongly agree). But research shows that such a ploy often backfires; that is, instead of preventing response bias, reverse item inclusion in a survey increases the risk of inattention and confusion. The summative score or the median score of all items should be the researcher team's sought-after quantitative measure. Consensus among researchers is that the summated score or the median score of summations is a much more valid and reliable measure of the construct under study than a single-item score. So, if the researcher wishes to report results on how well a population trusts its neighbors, the summative score or median score of summations should be reported along with a breakdown percentagewise of the choices[1] made by respondents on each item.

---

1   These choices are sometimes referred to as anchors. In a 1–5 scaling (e.g., 1 = strongly disagree, 2 = disagree, 3 = somewhat agree, 4 = agree, 5 = strongly agree), there are five anchors.

The position throughout this handbook is to accept that ordinal data are discrete values; as a result, nonparametric tests will be used to analyze ordinal data sets and Likert scaling data will be described using the discrete data point of a median score and illustrated with bar charts—as opposed to using central tendency measures of a mean score, standard deviation, and 95 percent confidence intervals that are commonly used to describe continuous, normally distributed data.

While it is beyond the scope of this handbook, there are methods of providing empirical evidence of the conceptual differences between anchors of Likert scaling. Equal distances between anchors could provide the researcher with more reliable and valid results.

COLLECTING ORDINAL DATA

It is also beyond the scope of this handbook to list all types of inventories, surveys, and indexes that collect ordinal data, yet a few examples are called for to clarify some points made. For example, when experts say that there is no standardized (equal) difference between ordinal data values, what we mean is that we have no evidence for saying the difference between two ordinal data values is the same for the distance between two other values on the same scale. Like the results of the third race at Belmont Park, it is possible for there to be unequal or unmeasurable differences between values—or horses. This fact prevents us from honestly calculating a mean or standard deviation for a set of ordinal data. For more clarity, let's look at another fictitious item in table 1.3, an item with somewhat ambiguous wording.

TABLE 1.3   Another Fictitious Likert-like Item Scoring

	Not True	Hardly True	Moderately True	True
Question A	1	2	3	4

It is the responsibility of the respondent to determine the difference between "hardly true" and "moderately true" within the context of the statement (item) being presented. As adverbs, the connotations of *hardly* and *moderately* are dependent on the reader's comprehension of the word and the context in which the adverbs appear. We cannot measure and compare them like we can snowfall or rainfall; nor can we say "hardly true" falls one standard deviation below the mean, as we have no mean as calculated in a normal distribution. Perhaps the respondent who chooses "hardly true" (a rank of 2) deems the statement in the item to be closer to "not true at all" rather than closer to "exactly true." Or perhaps the respondent finds the word *hardly* more concrete and definitive than the phrase "moderately true." Once the respondent commits to "hardly true," we accept the ranking of 2 for the purposes of measure. In short, we have made the descriptions temporarily irrelevant by converting them to ranks. It must be said here that this item could be highly improved on by using anchors of more clarity, such as these five anchors: never true, sometimes true, neither true nor false, often true, and always true. Or the researcher could use these four anchors as a set that excludes neutrality: never true, sometimes true, often true, and always true.

Let us now consider an item from the Behavioral Pain Scale (BPS; Payen et al., 2001), in which a nurse or health care professional completes the scoring based on observations. An excerpt of the BPS can be found in table 1.4.

TABLE 1.4   Facial Expression Component of the BPS

Item	Description	Score
Facial expression	Relaxed	1
	Partially tightened (e.g., brow lowering)	2
	Fully tightened (e.g., eyelid closing)	3
	Grimacing	4

Again, a description is converted into a rank. Also, we are taking for granted that a "relaxed" face is lower on the scale of pain than a "partially tightened" face. Some surveys come with very few descriptions, leaving the respondent to "fill in the blanks," so to speak. Consider an item (table 1.5) such as how a patient might receive their health care provider.

TABLE 1.5   Fictitious Health Care Rating Item

Using any number from 0 to 10, with 0 as the worst doctor possible and 10 as the best doctor possible, what number would you use to rate this doctor?
• 0 Worst doctor possible
• 1
• 2
• 3
• 4
• 5
• 6
• 7
• 8
• 9
• 10 Best doctor possible

There is no description in this item for such mistakes as "prescribed the wrong medicine" or "read the wrong chart," so it is up to the respondent to first qualify in their own mind what constitutes a "worst" or "best" doctor and then choose a rank up or down accordingly. This scaling does have an advantage over scaling that includes a choice such as "undecided," "neutral," or "unsure" in that subjects are rather forced to take a stance—and not given the option of neutrality.

For our last example, let us consider an item for which respondents are asked to provide the ranks next to the appropriate descriptions (table 1.6).

Jean-Francois Payen, Olivier Bru, Jean-Luc Bosson, Anna Lagrasta, Isabella Deschaux, Pierre Lavagne and Claude Jacquot, "Assessing Pain in Critically Ill Sedated Patients by Using a Behavioral Pain Scale," *Critical Care Medicine*, vol. 29, no. 12. Copyright © 2001 by Wolters Kluwer Health.

TABLE 1.6   Fictitious Item Asking Respondents to Rank Items

1. Rank in order of priority (1 = highest priority) aspects of a health plan for your family:
   - low deductible for primary care visitation
   - dental coverage
   - low monthly premium
   - prescription discount
   - eye care

We have just reviewed several different forms of survey items. Regardless of the form in which the survey items are written, it is important that the entire survey—that is, the collection of items—constitutes a valid and reliable tool. In other words, the survey must consistently measure what it purports to measure—whether it is measuring trust in one's neighbor, pain, or a doctor's efficiency. With that in mind, we will now look at validity and reliability in the context of a survey as a modern tool in today's research.

VALIDITY AND RELIABILITY IN COLLECTING ORDINAL DATA

If the researcher believes that the best method of answering their research question(s) is through the collection of ordinal data, it is essential that the researcher or research team use a collection tool (e.g., survey, inventory) that is measurably *valid* and *reliable*. There is evidence that can be reported to demonstrate the validity of a survey collecting ordinal data, such as correlations of scores with other surveys measuring the same construct; consequential evidence that the results of the survey led to proper placement of workers or proper treatment for therapy; the opinions of experts judging the survey; or interviews with those completing the survey to determine whether thinking processes properly matched the responses the respondents subsequently chose. Evidence of a tool's reliability can also be reported in various forms, such as a correlation value measuring the relatedness of two similar surveys taken by the same respondent over time (test-retest reliability) or, for Likert-like scales in particular, a coefficient alpha (Cronbach's alpha, which can be conducted in SPSS) measuring the consistency of the scores from the survey. Note that this list does not include all methods of evidence.

If no tool exists to measure a construct, the researcher or research team can develop and validate their own tool. But developing such a tool involves a meticulous process that is not outlined in this handbook. If a tool needs to be developed, it is strongly recommended that the researcher consult the most recent publication of *Standards for Educational and Psychological Testing* of the American Educational Research Association (AERA) or some other reputable guide to developing surveys collecting ordinal data or other types of data.

CONCLUSION

Ordinal data are valuable measures for researchers seeking to quantify and analyze sensory events or actual rankings of people or phenomena in the behavioral sciences and other fields of study. While the common practice of summating Likert-like values (treating ordinal data as though they were interval data) is discouraged by some members of the larger research community, such a practice can provide valuable insight into the human condition and worldly events. This practice of summating ordinal data (ranks) must be proceeded by the validation of the tool from which the ordinal data (ranks) are collected. Validity, reliability, administrative conditions, the mental and emotional state of the respondent, and the wording and phrasing within the items must all be highly scrutinized and reported alongside any results. Accordingly, researchers should consult the guidelines published by the AERA or other established, appropriate associations or publications to clarify any questions or concerns about the collection of ordinal data.

Exercises

1. Consider the descriptions used in the example. Discuss with a classmate whether any of the descriptors are questionable or ambiguous in creating the ordered ranks of 1–5.

	No Knowledge	Fair Knowledge	Average Knowledge	Much Knowledge	Great Knowledge
All items	1	2	3	4	5

2. For the researcher in the field of education surveying students or parents, what emotional or mental states or handicaps—even if temporary—on the part of the student as respondent might produce unreliable or invalid responses? In the field of health and wellness?
3. As a student, it is common to find a request for the completion of a course evaluation or campus facility evaluation in your academic email box. Can you recall a time when you failed to complete the survey? What were your reasons for nonresponse?
4. At the higher education level, students are almost always asked to evaluate the course and instructor on an end-of-semester basis. Too many requests can lead to *survey fatigue* on the part of the students—which in turn leads to nonresponses. Make some suggestions as to what colleges and universities can do to minimize or eliminate survey fatigue while maintaining an unbiased course and instructor evaluation process.
5. Discuss the advantages and disadvantages of collecting data of sensory events or emotions using Likert-like items compared to the qualitative methods of interviews and observations.

Distribution-Free Aspects of Nonparametric Tests

OBJECTIVES

- Understand parameters of the normal distribution.
- Apply the properties of the normal distribution to solve problems.
- Understand how ranks are used to determine probabilities.
- Apply knowledge of three distribution-free test statistics to solve problems.

KEY TERMS

Population mean (μ): The mean score of a population on some variable.

Standard deviation of a population (σ): A measure of how scores vary around a population mean.

Normality test: A probability test that determines the likelihood of a sample set(s) of data originating from a normal distribution.

Null hypothesis (H_0): The hypothesis that observed differences in scores between groups are due to chance.

Alternative hypothesis (H_1): The hypothesis that observed differences in scores between groups is not due to chance but rather due to some intervention or treatment.

Null distribution: The probability distribution of the test statistic when the null is true.

z-distribution: A normal distribution of standard scores with a mean of zero and a standard deviation of one.

A BRIEF INTRODUCTION TO THIS UNIT

In this unit, we will focus on two distributional shapes: (a) a parametric shape, in which the data are distributed along the normal (bell-shaped) curve or (b) a nonparametric shape, in which the data are not normally distributed. This handbook focuses on nonparametric data sets, of which ordinal data sets are practically always included. Once the researcher knows which shape the sample data set likely originates from (usually determined by the results of a Kolmogorov–Smirnov or Shapiro–Wilk test), they can proceed with choosing the appropriate hypothesis test, as seen in table 2.1.

TABLE 2.1   Hypothesis Tests According to Sample Data Distribution

Research Question/ One-Group Design	Parametric	Nonparametric**
Does a new treatment lead to scores different from known, established scores?	One-sample *t*-test	One-sample Wilcoxon signed rank test
Do scores change from one measure to another because of some intervention?	Paired samples *t*-test	Wilcoxon signed rank test
Do scores change over three or more measures because of some intervention?	Rep. measures ANOVA*	Friedman test
Research Question/2 + Group Design	Parametric	Nonparametric
Are scores different across two independent groups?	Ind. samples *t*-test	Wilcoxon rank sum test†
Are scores different across three or more independent groups?	One-way ANOVA	Kruskal-Wallis *H*-test
Does a significant relationship exist between two variables?	Pearson's coefficient of correlation	Spearman's rho or Kendall's tau

** All seven of these tests are described in this handbook.
* Repeated-measures analysis of variance.
† When conducting this test in SPSS, results of both this test and the Mann–Whitney *U*-test are given, as the tests are very similar.

REVIEW OF THE NORMAL DISTRIBUTION

Before we delve into distribution-free analyses, let us briefly review data that are normally distributed and not free—*not free* in the sense that the data adhere to the distribution parameters of a bell-shaped curve (normal distribution), and probability calculations are based on the normal distribution. First, let us assume graduate student age is normally distributed and that the mean (μ) age of graduate students in the US is 31 years with a standard deviation (σ) of two years. Using the parameters of the normal distribution, we can conclude that the probability of all US graduate students being between the ages of 29 and 33 years is .683. Extending the example further, we can conclude that roughly 99.7% of all graduate

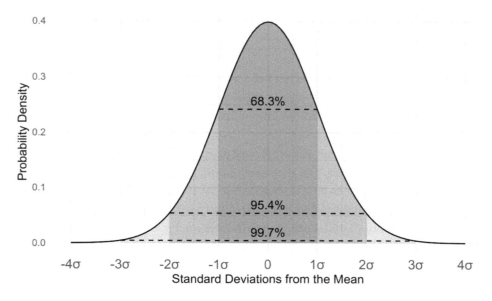

FIGURE 2.1.   Normal distribution.

Copyright © by D. Wells (CC BY-SA 4.0) at https://commons.wikimedia.org/wiki/File:Standard_Normal_Distribution.png.

students are between the ages of 25 (31 – 2 – 2 – 2 = 25) and 37 (31 + 2 + 2 + 2 = 37). We could also calculate the probability of a graduate student being 35 years old or older by converting 35 to a z-score:

$$z = (X - \mu) / \sigma$$
$$z = (35 - 31) / 2$$
$$z = 2.0$$

Table A (in the Tables section of this text) shows that when $z = 2.00$, the area beyond z is .0228. So, we can conclude that the probability of being a graduate student 35 years of age or older is 2.3 percent. This same calculation could have been made just by examining the normal distribution. The sum of the lightest gray areas above –2 σ and below –2 σ is (the two tails of the normal distribution) 1 – .954, or .046. Divide .046 by two (since we *only* want to know the area greater than $z = 2.0$ or the probability of students 35 years of age and older) and we get .023, or 2.3 percent.

Self-Check for Calculating *z*-Scores

What is the probability of a graduate student being 33 years or older?

Convert 33 to a z-score to determine the area beyond the z:

$$z = (33 - 31) / 2$$

$z = 1.0$. The area beyond 1.0 on the z-distribution (table A) shows that the probability of a graduate student being 33 years of age or older is .1587.

HOW RANKS ARE USED TO DETERMINE PROBABILITIES: THE *W* STATISTIC[1]

For ordinal data that are ranked, we do not need to ascribe to the underlying distribution of the population to determine probabilities such as those previously explained. We say we are distribution free with ranked ordinal data. Additionally, no knowledge of the mean and standard deviation is required.

For our first demonstration, assume we have ordinal pain scores, with no ties,[2] from two populations: a placebo group (X) and drug A group (Y). Note (table 2.1) that the pain scores have already been ranked and the ranks appear in parentheses. This example is just one of twenty different combinations (see table 2.2) of rankings that could be observed based on $m = 3$ and $n = 3$. The normal procedure in this case is to rank the lowest pain score a 1 and the highest pain score a 6. For further clarification, note that the first subject in the placebo group reported the highest pain score as evident by the ranking of (6); and the third subject in the drug A group reported the lowest pain score as evident by the ranking of (1).

TABLE 2.1 **Pain Scores by Group**

Placebo ($m = 3$)	Drug A ($n = 3$)
18 (6)	15 (4)
16 (5)	14 (3)
12 (2)	10 (1)

In table 2.2, we find all the possible combinations of rankings that Y, Y, and Y can be, given that the rankings for all six pain scores will range from 1 to 6 and we are only concerned with the possible combinations—not the order of those combinations. In other words, the (4, 3, 1) rankings will also represent the possibility of (3, 4, 1) or (1, 3, 4) or (1, 4, 3), as we will eventually attend to the sum of the rankings—a sum of eight in this case (1 + 3 + 4). The *W* represents the rank sum statistic. For our purposes, the *W* statistic will be used for null hypothesis testing with ordinal data as it is the appropriate statistic for the rank sum test.

---

1   The Mann–Whitney *U* statistic is an alternative to the *W* statistic when comparing the scores of two populations. They are essentially the same test, when $W = U + n(n + 1) / 2$.
2   Adjustments for ties will be demonstrated in subsequent units. SPSS automatically adjusts for ties in several analyses.

TABLE 2.2 Twenty Possible Combinations of Y, Y, Y Ranks with Randomly Chosen $m = 3$ and $n = 3$

Possible Y, Y, Y Combinations	Probability of This Combination Occurring	W (Sum of the Rankings)
1, 2, 3	1/20 or .05	6
1, 2, 4	.05	7
1, 2, 5	.05	8
1, 2, 6	.05	9
1, 3, 4	.05	8
1, 3, 5	.05	9
1, 3, 6	.05	10
1, 4, 5	.05	10
1, 4, 6	.05	11
1, 5, 6	.05	12
2, 3, 4	.05	9
2, 3, 5	.05	10
2, 3, 6	.05	11
3, 4, 5	.05	12
3, 4, 6	.05	13
4, 5, 2	.05	11
4, 5, 6	.05	15
6, 5, 1	.05	12
6, 5, 2	.05	13
6, 5, 3	.05	14

The probability of any one of the twenty combinations being observed is $1/20 = .05$.

Self-Check for Calculating a Probability

From a random selection of $m = 3$ and $n = 3$, what is the probability of observing a $W \geq 13$? We can use P_0 to denote the probability of a combination being observed.

It helps to first highlight in gray (table 2.2) all combinations of rankings when $W \geq 13$. We then add the probabilities (highlighted in gray) of all four combinations to determine the probability of observing $W \geq 13$ from a random selection of $m = 3$ and $n = 3$:

$$P_0(3, 4, 6) + P_0(4, 5, 6) + P_0(6, 5, 2) + P_0(6, 5, 3)$$
$$.05 + .05 + .05 + .05 = .20$$

So, the probability of observing $W \geq 13$ from these conditions is .20. We can confirm this by examining an excerpt from the upper-tail probabilities for the null distribution of the Wilcoxon rank sum W statistic (these n values are not included in table B of this text):

TABLE 2.3   Excerpt of the *W* Statistic Table when *m* = 3 and *n* = 3

		$n = 3$
x	*m* = **3**	*m* = **4**
11	.500	
12	.350	.571
13	.200	.429
14	.100	.314
15	.050	.200
16		.114

Source: Myles Hollander and Douglas A. Wolfe, *Nonparametric Statistical Methods*, p. 584.
Copyright © 1999 by John Wiley & Sons, Inc.

HOW RANKS ARE USED TO DETERMINE PROBABILITIES: THE *T*⁺ STATISTIC

Our second demonstration of how null distributions are calculated (i.e., how probabilities are formulated) is with the Wilcoxon signed rank T^+ distribution. Students might better recognize this statistic as the one used when we measure for a significant shift in attitude or level of agreement on a construct from pretreatment to posttreatment. For clarity of conceptual understanding, only three subjects will be considered, and their Likert scaling summated scores will be reported. Let us assume that the three randomly chosen subjects are asked to complete a ten-item survey measuring their levels of agreement on the use of e-cigarettes as a tobacco replacement among teens. An example of one item can be found in table 2.4.

TABLE 2.4   Sample Item from a Fictitious Ten-Item Survey Measuring E-Cigarette Safety for Teens

E-cigarettes are a safe alternative to smoking tobacco among teens ages 16–18.

Strongly Disagree	Disagree	Neither Disagree nor Agree	Agree	Strongly Agree
1	2	3	4	5

Immediately after reporting their level of agreement, the three subjects are shown a twenty-minute video on the potential harms to teens when using an e-cigarette. At the conclusion of the video, a ten-minute question-and-answer period takes place between the subjects and the presenter, with subjects then asked a second time to respond to the ten items in the survey. The results of the summated pretreatment and posttreatment scores, with no ties, can be found in table 2.5. Let (*D*) represent the difference between summated pretreatment and posttreatment scores, let *D* represent the absolute difference, and let B represent the *sum* of all the positive ranks. The lowest absolute difference value is given the lowest rank (1), and the highest absolute difference value is given the highest rank (3).

TABLE 2.5 **Pretreatment and Posttreatment Summated Scores and Other Values**

Subject	Pre-Sum	Post-Sum	Difference (D)	Absolute D	Rank	Code 0 = Negative 1 = Positive
1	46	44	−2	2	1	0
2	24	28	4	4	2	1
3	27	45	18	18	3	1
			B (number of positive ranks) =			2

Self-Check for Ranks

Of the three subjects involved in this fictitious study, two subjects demonstrated increased levels of agreement posttreatment, and one subject demonstrated a decreased level posttreatment. In sum, there are two positive ranks ($B = 2$), and the ranks of the two positive ranks are (2) and (3).

To calculate probabilities, we now need to consider all the possible positive ranks (B) and their rankings when three subjects are considered. Eight possibilities exist in such a case, as shown with their probabilities in table 2.6.

TABLE 2.6 **All Possible Outcomes with Three Subjects**

B (Number of Positive Ranks)	All possible Ranks Given B	Probability of Observing These Ranks	T^+ Sum of the Positive Tanks
0		1/8 or .125	0
1	$r_1 = 1$	.125	1
1	$r_2 = 2$	.125	2
1	$r_3 = 3$	.125	3
2	$r_1 = 1, r_2 = 2$	.125	3
2	$r_1 = 1, r_2 = 3$	.125	4
2	$r_1 = 2, r_2 = 3$	.125	5
3	$r_1 = 1, r_2 = 2, r_3 = 3$	.125	6

Self-Check for Positive Rankings

Note that B denotes the number of positive ranks for the purpose of calculating the probabilities of observing any combination of positive ranks. T^+ denotes the sum of those ranks for the purpose of calculating a p value to determine whether the difference from presurvey to postsurvey was significant:

> When $T^+ = 0$, none of the three subjects in the study exhibited an increase in agreement level from pre- to posttreatment.
>
> When $T^+ = 5$, two subjects in the study exhibited an increased agreement from pre- to posttreatment with rankings of **2** and **3**, as in tables 2.5 and 2.6.
>
> When $T^+ = 6$, all three subjects in the study exhibited an increase in agreement level from pre- to posttreatment.

Self-Check for Calculating a Probability

From a random selection of $n = 3$, what is the probability of observing a $T^+ \geq 5$? We can use P_0 to denote the probability of any possible T^+ being observed under such conditions.

In Table 2.7, all possible rankings whose sum is ≥ 5 are highlighted in gray. Again, we add the probabilities (highlighted in gray) of the two possible combinations of rankings to determine the probability of observing $T^+ \geq 5$ from a random selection of $n = 3$:

$$P_0(2, 3) + P_0(1, 2, 3)$$
$$.125 + .125 = .250$$

TABLE 2.7 **All Possible Outcomes with Three Subjects**

B (Number of Positive Zs)	All Possible Rankings Given B	Probability of Observing These Rankings	T^+ Sum of the Positive Rankings
0		1/8 or .125	0
1	$r_1 = 1$	.125	1
1	$r_2 = 2$	.125	2
1	$r_3 = 3$	.125	3
2	$r_1 = 1, r_2 = 2$	.125	3
2	$r_1 = 1, r_2 = 3$	.125	4
2	$r_1 = 2, r_2 = 3$	.125	5
3	$r_1 = 1, r_2 = 2, r_3 = 3$	.125	6

From here, we can confirm our calculation by consulting an excerpt (table C) of the selected upper-tail probabilities of the Wilcoxon signed rank T^+ found in the Tables section of this text.

TABLE 2.8   Excerpt of the Table when $n = 3$

	x	$P_0\{T^+ \geq x\}$
$n = 3$	3	.625
	4	.375
	5	.250
	6	.125

Source: Myles Hollander and Douglas A. Wolfe, "When n = 3," *Nonparametric Statistical Methods*, p. 576. Copyright © 1999 by John Wiley & Sons, Inc.

HOW RANKS ARE USED TO DETERMINE PROBABILITIES: THE *H* STATISTIC

The third and final demonstration entails the selected upper-tail probabilities for the null distribution of the Kruskal–Wallis *H* statistic. Let us assume that student athletes competing in countywide 200 m races receive one of three training methods for six consecutive weeks. Two athletes are randomly chosen from each of the three method groups (see table 2.9). The numbers 1–6 represent the rankings (placement) of the six athletes in a recently ran 200 m race with no ties.

TABLE 2.9   Rankings of Six Athletes and Their Placements in a 200 m Race by Training Method

Method A	Method B	Method C
(1)	(5)	(4)
(2)	(3)	(6)

There are ninety possible rank assignments with three groups ($k = 3$), and $n_1 = 2$, $n_2 = 2$ and $n_3 = 2$, so for the sake of space, those combinations will not be listed here. We can thank statisticians who have done the work for us by creating the probability table for the *H* statistic when the null is true. Table 2.10 is an excerpt of the null distribution for the *H* test below (table D) when comparing three groups, with two subjects in each group.

TABLE 2.10   Excerpt of the *H* Statistic Table when $k = 3$ and All $n = 2$

n_1	n_2	n_3	x	$P_0\{H \geq x)$
			$k = 3$	
2	2	2	3.714	.200
			4.571	.067

Source: Myles Hollander and Douglas A. Wolfe, "H Statistic Table when k = 3 and all n = 2," *Nonparametric Statistical Methods*, p. 634. Copyright © 1999 by John Wiley & Sons, Inc.

In this situation, we would need a calculated H value[3] of 4.571 or greater to be able to reject the null hypothesis at the .067 level[4] (likelihood of making a type I error). Even if such a rejection was made based on the results, we could only conclude that there is at least one method group significantly different than another. We can refer to this initial test as the *omnibus test*; post hoc analyses would then be required to determine which training method is superior to the other(s).

DISTRIBUTION-FREE TESTS AS ALTERNATIVES TO THE INDEPENDENT SAMPLES t-TEST, PAIRED SAMPLES t-TEST, AND ONE-WAY ANOVA

The distribution-free test statistics covered in this unit are also useful when working with data that fail to meet assumptions required for other tests, such as when the test variable is not normally distributed or when variances between groups are unequal. When the assumptions of either parametric or nonparametric tests are met, proceed with the test. Clearly, if the data are ordinal, use a distribution-free analysis. Such analyses are quite reliable, even though in the process of calculation raw scores are converted to ranks.

CONCLUSION

When a test variable is normally distributed, we can use the standard normal distribution (z-distribution) to calculate probabilities and to conduct parametric tests. The researcher analyzing ordinal data or nonparametric data has the option of using several distribution-free tests that are easy to understand and that provide various statistical outcomes—especially when it comes to testing the null hypothesis. These tests are also available through SPSS, with software that automatically adjusts for ties when applicable.

Exercises

1. Twenty randomly selected school staff members are asked to rate the leadership effectiveness of the school's administration. Half of the subjects are faculty members, and half are nonfaculty members. The regional director wants to know if rating scores are different for the ten faculty members and ten nonfaculty members (e.g., clerical, janitorial, nursing staff). The regional director uses $\alpha = .053$. She hand-calculates a W statistic of 121 ($x = 121$), $n = 10$, and $m = 10$. Should she reject the null hypothesis that the scores are the same for both groups? Use the values found in table B as a guide.

---

3 Calculation not shown here. That calculation will be shown in unit 5.
4 In the behavioral sciences, $\alpha = .05$ is the most common level used for rejection of the null, although .067 is acceptable if the researcher wishes to take a risk of making a type I error at this level.

2. A home appliance service manager asks thirty customers to rate a new serviceman using the following item and scaling:

How satisfied were you with the on-the-job cleanliness of the serviceman on his recent visit to your home?

Extremely Unsatisfied	Unsatisfied	Somewhat Unsatisfied	Neutral	Somewhat Satisfied	Satisfied	Extremely Satisfied
1	2	3	4	5	6	7

The results of the survey are not favorable for the serviceman. The serviceman is then asked to complete an eight-hour course in on-the-job cleanliness, which he completes. The manager then asks thirty more customers to rate the serviceman, using a matched-pairs design whereby the thirty new customers are like the first thirty customers on such variables as age, gender, square feet of home, minutes on the job, and type of appliance repaired. The manager uses $\alpha = .0502$ and hand-calculates a T^+ value of 314 ($x = 314$). Can the manager be confident that the serviceman improved on his on-the-job cleanliness? Use the values found in table C as your guide.

3. A hospital director wants to know if a new treatment designed to decrease the pain levels of patients suffering from rheumatoid arthritis is more effective than the two different treatments currently being administered to these patients.[5] Twenty-four patients currently receiving treatment at the hospital are randomly selected and then randomly assigned to one of the three treatments, with eight patients in each group. Twenty-four hours after administration of the treatments, the patients self-report hand pain levels using ordinal data scaling. The director uses $\alpha = .0101$. He hand-calculates an H value of 7.000. Is there any statistical evidence of significant difference in hand pain between the three groups? Refer to table D for H statistic values.

---

5   This procedure would normally be done as a phase III trial, possibly involving thousands of subjects, but for the purposes of this exercise, a much lower number of subjects is being used.

The Rank Sum Test for Two Independent Samples

OBJECTIVES

- Understand and apply procedures of the Wilcoxon rank sum test.

- Understand procedures for correcting for ties.

- Analyze and create a results section for the rank sum test.

- Conduct a Wilcoxon rank sum test using SPSS.

KEY SYMBOLS[1]

SCR_X: Scores for the X-group

SCR_Y: Scores for the Y-group

W statistic: Sum of the ranks for either the Y-group or X-group

x (W_{cv}): the value corresponding to a particular n and α when the null hypothesis is true

V_0: the variance of W when the null is true, given m and n

KEY TERMS

Type I error: Rejecting the null hypothesis when the null hypothesis is true.[2]

Type II error: Failing to reject the null hypothesis when the null hypothesis is false.

---

1   These are arbitrary symbols used as teaching tools in this handbook. There are other symbols used for the same concepts in various other textbooks.
2   A type I error is sometimes referred to as a false positive; a type II error is sometimes referred to as a false negative.

A BRIEF INTRODUCTION TO THIS UNIT

The use of the equations in this unit for calculating the W statistic[3] serves as a teaching tool, designed to support conceptual understanding of the test itself. Since the advent of computers, most researchers rely on computer software to do the calculations, which will be practiced at the end of this unit.

When reference is made to either upper-tail probabilities or the area beyond z, the corresponding value reflects the p value of a one-tailed test. SPSS provides two p values for the rank sum test:

- An asymptotic (approximation of the normal distribution) two-tailed p value, shown as Asymp. Sig. (2-tailed).

- An exact p value, shown as Exact Sig. [2*(1-tailed) Sig.)]. This value is the product of the exact one-tailed p value times two.

So, if the researcher's alternative hypothesis is that scores or ranks on some outcome measure will increase and they propose, a priori, a one-tailed test at $\alpha = .05$, they can divide the exact significance value by two for the one-tailed p value for determining whether to reject the null hypothesis and the likelihood of making a type I error.

The asymptotic p value reported by SPSS is adjusted for ties. The exact p value is not. Hence, it is recommended that, when no ties or few ties exist, the researcher rely on the exact p value. When there are multiple ties in the data set, it is wise to rely on SPSS's asymptotic p value.

SETTING ALPHA A PRIORI

It is standard practice to set alpha before conducting any statistical analyses. The basis for setting alpha (level of significance) might be based on prior research results, information found in a literature review, or after carefully considering the consequences of making a type I or type II error. While $\alpha = .05$ is commonly used, there is nothing magical about the value. It is merely a value that may ultimately provide[4] a certain level of comfort or confidence. For example, when the risks are low (minimal negative consequences), it may behoove the researcher to set alpha at .10; broader parameters have now been set for rejection of the null,[5] which might be the impetus for a new practice or treatment that ends up being highly effective.

---

3   The W statistic is a product of the Wilcoxon rank sum test; the U statistic is a product of the Mann–Whitney U test, which is another version of the rank sum test. Both statistics are provided by SPSS, and SPSS provides the results of both tests.
4   We often don't know whether we made a type I or type II error until others replicate the study and more evidence is accumulated.
5   While moving alpha from .05 to .10 decreases the likelihood of a type II error (failing to reject the null when the null is false), it increases the likelihood of making a type I error (false positive).

PROCEDURES FOR CONDUCTING A TWO-TAILED TEST FOR SMALLER SAMPLE SIZES

There are often situations in which a researcher wishes to compare scores of two independent groups while testing the alternative hypothesis that the scores are not equal after some treatment or intervention has taken place. This is called a *two-tailed test* or *nondirectional test*. For the purposes of this unit, we will use the acronym SCR for scores, X for the scores of one group, and Y for the scores of the other group. The null hypothesis will be represented by $SCR_Y = SCR_X$, and the alternative hypothesis represented by $SCR_X \neq SCR_Y$.

The Research Problem Testing H_0: $SCR_Y = SCR_X$

A member of Smalltown Chamber of Commerce investigates whether citizens favor one pizzeria over the other, a task made easier by the fact that there are only two pizzerias in town. Six patrons are randomly selected to rate the two establishments on the restaurants' cleanliness, in-house service, online service, drinks, food options, and food quality. Likert-like scores (ordinal data) from the survey were summated, with the highest summated score of 60 indicating the highest possible top rating.

Step 1: Verify That Assumptions for the Test Are Met

Assumption 1: The one dependent variable is at least ordinal in nature.
Assumption 2: The one independent variable consists of two independent groups.
Assumption 3: The distribution of scores in both groups are similar in shape.[6]

Step 2: State the Hypotheses

The null hypothesis is that there is no difference in the rankings of the two pizzerias:

$$H_0: SCR_Y = SCR_X$$

The alternative hypothesis is that the rankings of the pizzerias will be significantly different:

$$H_1: SCR_Y \neq SCR_X$$

Step 3: Set the Critical Value for Rejection of the Null Hypothesis

We will use X ($m = 6$) to represent the rankings for Tony's Pizza and Y ($n = 6$) to represent the rankings for Josie's Pizzeria. Using the upper-tail probabilities for the null distribution of the Wilcoxon rank sum W statistic (table B), when $m = 6$ and $n = 6$, with $\alpha = .047$[7], x (W_{cv}) = 50. Since this is a two-tailed

---

6   Violation of this assumption means that rejection of the null is an indicator that the scores are significantly different but that the median scores are not necessarily different.
7   This is an arbitrary decision, made by the researcher when determining at what level they are willing to risk making a type I error based on the availability in the table.

test, we need to determine whether one group ranks higher than the other (upper tail) or if one group ranks lower than the other (lower tail).

Upper tail: Reject the null hypothesis if $W \geq x\ (W_{cv})$.
Upper tail: Reject the null hypothesis if $W \geq 50$.
Lower tail: Reject the null hypothesis if $W \leq n(n + m + 1) - x\ (W_{cv})$.
Lower tail: Reject the null hypothesis if $W \leq 6(6 + 6 + 1) - 50$.
Lower tail: Reject the null hypothesis if $W \leq 28$.

Self-Check for Setting Alpha

An examination of the table (see table 3.1) with $m = 6$ and $n = 6$, with $\alpha = .047$ shows an $x\ (W_{cv})$ value of 50. The choice of .047 was arbitrary: The researcher could have easily chosen .021, thereby reducing the chance of making a type I error down to 2.1 percent. If such were the case, the researcher would then need to use a W statistic of 52 when making their calculations regarding rejection of the null hypothesis.

TABLE 3.1 **Excerpt of the Upper-Tail Probabilities for the Null Distribution of the W Statistic**

	n = 6
x	m = 6
39	.531
40	.469
41	.409
42	.350
43	.294
44	.242
45	.197
46	.155
47	.120
48	.090
49	.066
50	.047
51	.032
52	.021
53	.013
54	.008
55	.004
56	.002
57	.001

Source: Myles Hollander and Douglas A. Wolfe, "Upper Tail Probabilities for the Null Distribution of the W Statistic," *Nonparametric Statistical Methods*, p. 588. Copyright © 1999 by John Wiley & Sons, Inc.

Step 4: Calculate *W*

Now that we have established the criteria for rejecting the null hypothesis we need to calculate the observed *W* statistic from the data set.

Upper tail: Reject the null hypothesis if W ≥ 50.
Lower tail: Reject the null hypothesis if W ≤ 28.

As seen in table 3.2, the process is straightforward. With no ties in ranks, assign the rank of (1) to the lowest ranking and a rank of (12) to the highest ranking across the two pizzerias. Then add the ranks for each column for the observed *W* statistic.

TABLE 3.2   Survey Scores and Ranks in Parentheses of Tony's Pizza and Josie's Pizzeria

Tony's Pizza (X, *m* = 6)	Josie's Pizzeria (Y, *n* = 6)
35 (12)	33 (11)
31 (10)	30 (9)
26 (6)	29 (8)
27 (7)	24 (4)
25 (5)	23 (3)
20 (1)	21 (2)
W = 41	W = 37

Self-Check for Rankings

There is a method of self-checking for proper rankings using the following formula:

$N(N + 1) / 2 = W_x + W_y$, when N is the total number of subjects	
12 (12 + 1) / 2 = 78	41 + 37 = 78

At this point, we should feel good about our rankings.

Suggestion about Data Layout

If you are laying out data values by hand on a sheet of paper, it is recommended that you enter the values of the group with the smallest sample size in the right column. For example, if we have only five rankings for Tony's Pizza and six rankings for Josie's Pizzeria, place the rankings of Tony's Pizza in the right column, so *m* = 6 and *n* = 5. It is the author's opinion that this method facilitates the use of the table for selected upper-tail probabilities for the null distribution of the Wilcoxon rank sum *W* statistic.

Step 5: Interpret the Results

It is suggested to use the smaller *W* value of the two as the statistic in determining whether to reject the null hypothesis. SPSS output provides the smaller value as the Wilcoxon *W*: *W* = 37.

Upper tail: Reject the null hypothesis if $W \geq 50$.

Lower tail: Reject the null hypothesis if $W \leq 28$.

Upper tail: Reject the null hypothesis if $37 \geq 50$. The conclusion is to fail to reject the null hypothesis.

Lower tail: Reject the null hypothesis if $37 \leq 28$. The conclusion is to fail to reject the null hypothesis.

We lack evidence that the pizzerias differ in rankings when $\alpha = .047$.

PROCEDURES FOR CONDUCTING A TWO-TAILED TEST FOR LARGER SAMPLE SIZES

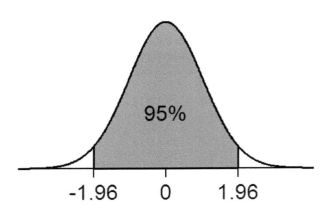

FIGURE 3.1.   Cut-off values for a two-tailed test.

Copyright © by Qwfp (CC BY-SA 3.0) at https://commons.wiki-media.org/wiki/File:NormalDist1.96.png#mw-jump-to-license.

The table of upper-tail probabilities for the null distribution of the Wilcoxon rank sum W statistic accommodates up to $n = 10$ and $m = 10$, but for larger sample sizes we have a standardized formula that produces an approximate z value.[8] With a z value, we can use the parameters of the normal distribution for criteria for rejecting the null hypothesis as a two-tailed test (figure 3.1).

We will use the same W score of 37 from the pizzeria example to demonstrate use of the formula for larger sample sizes. Again, since the alternative hypothesis is that ranking of the two pizzerias are different, we will use the two extreme ends of the normal distribution when each tail contains .025 of the distribution. The z values corresponding to the two ends are 1.96 and –1.96, which means we should

reject the null hypothesis when z > than 1.96 or z < –1.96.

We will use the following formula to calculate z for the Wilcoxon rank sum test:

$$W - MW_0 / V_0$$

Calculation alert! The square root sign in this formula is designated by superscript .5 as seen outside and to the upper-right of the brackets. Also, first subtract 37 – 39 before dividing (adjusting) the outcome by the variance. In this case, do not follow the rules of PEMDAS.

We know that $W = 37$. The formula for MW_0 is $n (m + n + 1) / 2$, and the formula for V_0 is

$$[mn (m + n + 1) / 12]^{.5}.$$

$$MW_0 = n (m + n + 1) / 2 = 6(13) / 2 = 39$$
$$V_0 = [mn (m + n + 1) / 12]^{.5} = [36(13) / 12]^{.5} = 6.245$$
$$37 - 39 / 6.245 = -.320, \text{ so } z = -.320$$

---

8   This formula will produce an approximate z value that will slightly differ from the Exact Sig. [2*(1-tailed Sig.)] p value that SPSS provides. Since there are no tied ranks in the pizzeria exercise, the Exact Sig. [2*91-tailed Sig.)] p value of .818 should be used in deciding whether to reject the null.

Since –.320 is not to the left (area of rejection) of –1.96, we fail to reject the null hypothesis. We have no evidence that the pizzerias differ in rankings. Referring to table A, we see that when $z = .320$, $p = .3745$.[9]

Correcting for Ties

When correcting for ties, the V_0 is reduced. Think about it: Since variance is a measure of positive and negative deviations, ties are evidence that the variance has been reduced—the deviations are not as broad. To correct for ties, we will use a formula based on the number of tied values and reduce the variance accordingly. First imagine that there were four tied scores in the pizzeria survey (table 3.3).

TABLE 3.3 **Survey Scores and Rankings in Parentheses of Tony's Pizza and Josie's Pizzeria**

Tony's Pizza (X, $n = 6$)	Josie's Pizzeria (Y, $m = 6$)
35 (12)	33 (11)
31 (10)	30 (8.5)
26 (6)	30 (8.5)
27 (7)	24 (4)
25 (5)	22 (2.5)
20 (1)	22 (2.5)
W = 41	W = 37

The values of 22 and 22 share the second and third ranking; hence, their tied rank is 2.5 (as 2 + 3 / 2 = 2.5). The same goes for the tied values of 30 and 30. They share the eighth and ninth ranks; hence, their tied rank is 8.5 (8 + 9 / 2 = 8.5). We will use the following formula to adjust for the tied ranks: $t^3 - t / 48$, with t representing the number of tied ranks. Calculation alert: The square root sign in this formula is designated by superscript .5, as seen outside and to the upper-right of the brackets.

Using the ties in table 3.3, $4^3 - 4 / 48 = 1.25$. To adjust the variance, we subtract 1.25 from the calculated value of the variance just before squaring it:

$$\text{adjusted } V_0 = \{[36(13) / 12\} - 1.25]^{.5}$$
$$\text{adjusted } V_0 = [39 - 1.25]^{.5}$$
$$\text{adjusted } V_0 = 6.144$$

Now when we use adjust V_0, the calculation for z is 37 – 39 / 6.144 = –.325. So, adjusted $z = .325$.

Self-Check for Understanding the New Values of z and p

With four tied values, the V_0 was reduced from 6.245 to 6.144. We expect a reduction in the variance with ties since there are fewer deviations. The z value changed from –.320 to –.325, making it less likely that we reject the null hypothesis. This should also be somewhat expected, as ties are evidence of less

---

9 This is a one-tailed p value, as we referenced the upper-tail probabilities; SPSS provides a two-tailed p value of .749. The p value of .3745 as calculated is exactly half of .749 as reported by SPSS.

difference in the two variables. With that said, we can also see that adjustments for ties make little difference in our decision making of whether to reject the null hypothesis.[10] For the remainder of this handbook, we will rely on SPSS to correct for ties when applicable with one exception in unit 5 when we will compare our tie-adjusted calculations to the tie-adjusted calculations of SPSS.

CALCULATING A REQUIRED SAMPLE SIZE

Several methods of calculating a sample size are available: the use of various tables, various formulas, and software programs, the most popular being G*Power. All three methods require that the researcher (a) establish the desired power[11] of the test $(1 - \beta)$, often set at .80 or .90; (b) the directional nature of the test, one-tailed or two-tailed; (c) alpha a priori, often set at .05[12] in the behavioral sciences; (d) the effect size[13]; and (e) the ratio of group sizes (N2/N1). If the researcher has access to G*Power, the process is rather straightforward. For demonstration of G*Power, let us assume that the researcher hypothesizes that a new treatment will lead to greater levels of confidence among females compared to males. The researcher wants to be at least 95 percent confident in the results ($\alpha = .05$), have an 80 percent chance of rejecting the null hypothesis if the null is false, use a one-tailed test (as the alternative hypothesis is that greater confidence among girls), and is searching for a large effect size of .80. Consider the ratio to be 1 (N2/N1). The test family should be set at *t tests*, and the statistical test set at *Means: Wilcoxon-Mann-Whitney test (two groups)*. According to the calculations of G*Power, at least 21 subjects are needed in each group to reach the actual power of .80003744 for this rank sum test. With that said, the researcher should strive to include as many subjects as possible, for power is increased when sample size is increased.

EFFECT SIZE AND CLINICAL AND STATISTICAL SIGNIFICANCE

An effect size tells us how meaningful a difference is between groups on a particular variable. Knowing the effect size is important to researchers. Consider that researchers might find a *statistically significant* difference in pain reduction between two treatments, with treatment B leading to significantly lower pain scores than treatment A with a certain type of patient, but with a very small effect size (as made evident by the appropriate effect size statistic). It could be that the cost of administering treatment B outweighs the reduction of pain provided by treatment B. In other words, there is no *clinical significance*, as the difference in pain between the two treatments—despite being statistically significant—is

---

10 When there are many ties, decisions about the null hypothesis could be affected.
11 Power refers to the ability to reject the null hypothesis when the null is false.
12 A much lower alpha, such as .001, might be used in the medical field when the consequences of being wrong (a type I error) could lead to patients being harmed due to some unproven treatment or drug.
13 The effect size is defined as the degree to which the phenomenon under study exists.

negligible and not worth the efforts of changing drugs, procedures, or policy. Of course, a different situation could occur: A researcher might find both statistical and clinical significance in the results of a study. Ultimately, it is the responsibility of stakeholders to determine whether an effect size is large enough to warrant changes in treatment, procedures, or policies. It is also common for researchers to settle on the desired or required effect size prior to conducting the study, which helps to facilitate the decision-making process once the statistical results are in.

There are methods of calculating the effect size for nonparametric statistical analyses. Those procedures are not covered in this book. It is recommended that students consult their instructor or turn to published formulas and available statistical software for determining an effect size for the analysis. This is also the case for several estimators and confidence intervals. They can be calculated for nonparametric statistical analyses, but the procedures for those calculations are not covered by this author at this time. Finally, it is suggested that you report the raw difference found in your analysis for future researchers to consider.

CONDUCTING A RANK SUM TEST IN SPSS

The first step in conducting a Wilcoxon–Mann–Whitney rank sum test in SPSS is to define the variables. For demonstration purposes, we will use the same data (table 3.2) used for the pizzeria study without ties. After opening a new data field in SPSS, click on the *Variable View* tab in the lower-left corner of the field. Create two variables for the rank sum test: *group* and *scores*. It is suggested that, after first entering *group* and *score*, press the Tab button on the keyboard to preload the data fields (see table 3.4).

TABLE 3.4 **Variable View in SPSS Data Field**

	Name	Type	Width	Decimals	Label	Values	Missing	Columns	Align	Measure	Role
					Untitled 1 [Data Set0] – IBM SPSS Statistics Data Editor						
1	group	numerical	8	0	group	1, Tony's	none	8	right	nominal	input
2	score	numerical	8	0	score	none	none	8	right	ordinal	input
3											

For *group*, we need no decimals (optional). Under Name, type in a label such as *group* or *pizza group*. In the Values box to the right of the word *none*, click on the blue icon (it is there but hidden) to activate the *Value Labels* box. Enter 1 and label it Tony's Pizza; enter 2 and label it Josie's Pizzeria. Click OK. Drag down the Measure box to make group *nominal* because group names are nominal data.

For Name, type in *score*; then enter *score* or *survey score* for Label and drag down Measure to make score *ordinal data*. It is optional to change Decimals to 0, but I recommend it as a habit as nominal data are discrete and technically have no values other than the ranks they are assigned.

Now click the *Data View* tab in the lower-left corner of the data field. This is when we enter the data for the rank sum test (table 3.5).

TABLE 3.5   Data View in SPSS Data Field

	group	score	
1	1	35	
2	1	31	
3	1	26	
4	1	27	
5	1	25	
6	1	20	
7	2	33	
8	2	30	
9	2	29	
10	2	24	
11	2	23	
12	2	21	
13			

Untitled 1 [Data Set0] – IBM SPSS Statistics Data Editor

Source: Generated with SPSS software. Copyright © 2023 by IBM Corporation. Reprinted with permission.

For subjects 1–6 ranking Tony's Pizza, enter 1. For subjects 7–12 ranking Josie's Pizzeria, enter 2. Then enter the corresponding summated scores for each subject for each pizzeria. Once the data are entered, the drag-down process from the top of the data field is as follows: ANALYZE > NONPARAMETRIC TESTS > LEGACY DIALOGS > 2 INDEPENDENT SAMPLES. At this point, you are provided with the test field. Highlight *group* and move it inside the Grouping Variable box. Click on Define Groups and label Group 1 a *1* and Group 2 a *2*. Click Continue. Then highlight *score* and move it inside the Test Variable List box. Click OK.

The results are in table 3.5. Note that although Josie's Pizzeria achieved a higher mean rank (6.83) than Tony's Pizza (6.17), the differences are considered insignificant. We will examine the significance of the two-tailed test (in this case, .749) for deciding whether to reject the null hypothesis. As with earlier calculations, we fail to reject the null hypothesis, $p = .749$.

TABLE 3.6   Modified Results Section of Wilcoxon–Mann–Whitney Rank Sum Test

Ranks				
Pizza group		N	Mean rank	Sum of ranks
Survey score	Tony's	6	6.83	41.00
	Josie's	6	6.17	37.00
	Total	12		

Test Statistics[a]	
Mann–Whitney U	16.000
Wilcoxon W	37.000
Z	−.320
Asymp. Sig. (2-tailed)	.749
Exact. Sig. [2*(1-tailed sig.)]	.818[b]

a. Grouping variable: Pizza group
b. Not corrected for ties

Source: Generated with SPSS software. Copyright © 2023 by IBM Corporation. Reprinted with permission.

CREATING AN APA RESULTS SECTION FOR THE RANK SUM TEST

We will use the data from the SPSS results to report our findings. While the *Publication Manual of the American Psychological Association* (APA, 2020) does not provide a specific formatting guide for the rank sum test, we will use this guide for statistical abbreviations and symbols:

> A Wilcoxon rank sum test[14] was conducted to evaluate the null hypothesis that no differences exist in the rankings of two pizzerias. The test was not significant, $z = -.320$, $p = .818$. No significant difference was found between the rankings of the two pizza parlors.

---

14   Since SPSS provides both the *W* statistic and the Mann–Whitney *U*, the researcher could state that a Wilcoxon–Mann–Whitney rank sum test was conducted.

CONCLUSION

Researchers can use the rank sum test to measure for significant differences in scores between two independent groups when the data are ordinal, ranked, or non-normal. The test is quite efficient and provides approximate p values. Tables are available for smaller sample sizes, and a formula (standardized) is available for larger sample sizes. Adjustments can be made for ties, and estimated required sample sizes can be calculated using various methods, including G*Power.

Exercises

1. Consider the survey scores and rankings for the pizzeria study; we calculated $W = 41$ and $W = 37$ with a difference of four between the two sum of ranks. Given that there are 12 scores, what is the greatest difference between the sum of ranks that could occur? What would be the values of those W statistics? Hint: You do not need to use any scores for this exercise but simply separate and plot the 12 ranks differently than they are in table 3.2.

2. In your own words, explain the effect of ties on (a) the variance when the null hypothesis is true and (b) the approximate z value.

3. Discount Tire Company developed a new tire designed to better handle inclement weather. The Research & Development Division randomly chose seven new and seven old tires for a test of any difference in mileage before complete tire wear down between the two designs. Test the null hypothesis that there is no difference in mileage between the old and newly designed tires. Assume that the data are non-normal. Conduct a two-tailed test using the smaller sample size calculations with $\alpha = .049$. Rank the values and follow the procedures for conducting a two-tailed test for smaller samples. Use the following data set and refer to table B for the W statistics.

Old Design Tires	Mileage	Rank	Newly Designed Tires	Mileage	Rank
1	42,000		1	38,000	
2	41,500		2	37,500	
3	41,000		3	37,000	
4	40,500		4	36,500	
5	40,000		5	43,000	
6	39,500		6	36,000	
7	39,000		7	35,500	
W =			W =		

4. A sociologist wants to know whether younger adults demonstrate greater concern over the effects of climate change than older adults. Using Likert scaling, she randomly surveys ten Gen Z college freshmen and ten baby boomers. A higher score indicates greater concern. Test the null hypothesis that there is no difference in concern over the effects of climate change between the two generations. Consider the data ordinal. Conduct a two-tailed test using the formula for larger sample size with $\alpha = .05$. State whether you reject the null hypothesis or fail to reject the null. Use the following data set.

Gen Z Subjects	Summated Concern Score	Rank	Baby Boomer Subjects	Summated Concern Score	Rank
1	58		1	15	
2	42		2	31	
3	40		3	32	
4	38		4	18	
5	34		5	17	
6	36		6	25	
7	30		7	16	
8	29		8	14	
9	19		9	12	
10	13		10	11	
W =			W =		

5. An educator is interested in the mathematics achievement of third graders taught under two diffe-rent conditions. The first condition includes the use of modern, standard classroom equipment, such as a smart board, learning touch pads, textbooks, workbooks, and other standard classroom equipment. The second condition includes all the equipment found in the first condition but also includes colored manipulatives. Using SPSS, test the null hypothesis that no differences in math achievement scores exist between the two conditions. Assume that students were randomly selected and randomly placed into two conditions and that the data are non-normal. Set $\alpha = .05$. Write an APA results section as your answer. Use the data set that follows. For this exercise, you can ignore the household income rank, which is included for use in exercise 3 in unit 6.

	Condition 1			Condition 2	
Student	Achievement Score	Household Income Rank	Student	Achievement Score	Household Income Rank
1	35	2	1	52	2
2	51	3	2	87	3
3	66	3	3	76	3
4	42	4	4	62	1
5	37	2	5	81	2
6	46	4	6	71	2

(*Continued*)

| | Condition 1 | | | Condition 2 | |
Student	Achievement Score	Household Income Rank	Student	Achievement Score	Household Income Rank
7	60	1	7	55	2
8	55	3	8	67	4
9	53	3	9	77	3
10	68	2	10	78	3
11	70	4	11	80	4
12	80	2	12	82	2
13	80	2	13	70	4
14	76	3	14	71	2
15	75	3	15	72	3
16	74	3			
17	66	4			

6. This exercise was developed from data shared by the National Aeronautics and Space Administration (NASA). The answers to this exercise are *not* included in the Answers to Unit Exercises section of this text. You may want to consult your instructor for assistance or for confirmation of your results. You are also free to conduct the test either by hand or by using SPSS.

Following are twenty meteorites randomly chosen from a list of 34,513 meteorites known to have fallen between the years 1769–2013. With this subsample, do we have any evidence of significant differences in mass (g) between the two independent groups of meteorites? Set $\alpha = .05$ if using SPSS. Set $\alpha = .053$ if conducting the test by hand.

| | Fell between 1823–1831 | | | Fell between 2002–2013 | |
Year	Name	Mass (G)	Year	Name	Mass (G)
1823	Botschetschki	614	2002	San Michele	237
1823	Nobleborough	2,300	2006	Bassikounou	29,560
1823	Allan Hills A78210	8.95	2006	Moss	3,763
1824	Renazzo	1,000	2007	Mahadevpur	70,500
1824	Tounkin	2,000	2009	Ash Creek	9,500
1824	Zebrak	2,000	2010	Lorton	329.7
1825	Honolulu	2,420	2011	Boumdeid	3,599
1825	Nanjemoy	7,500	2011	Tissint	7,000
1828	Richmond	1,800	2012	Battle Mountain	2900
1831	Wessely	3,750	2013	Chelyabinsk	100,000

The Signed Rank Test for Paired Samples

OBJECTIVES

- Understand and apply procedures of the Wilcoxon signed rank test.

- Understand and apply procedures for correcting for ties.

- Understand and apply the procedures for a two-sided test.

- Analyze and create a results section for the signed rank test.

- Conduct a Wilcoxon signed rank test using SPSS.

KEY SYMBOLS[1]

C: Change in scores or ranks

$\mathbf{x}$ ($\mathbf{C_{cv}}$): The value corresponding to a particular n and α when the null hypothesis is true

(D): Difference in scores/ranks

D: Absolute value of the difference

T^+: The sum of the positive ranks

$\mathbf{V_0}$: Variance when the null is true

---

1   The symbols used for this unit are being used arbitrarily by the author; the same concepts may be represented by different symbols in other textbooks.

Lower-tail test: A test measuring for a decrease in a test variable across time or treatment(s).

Upper tail test: A test measuring for an increase in a test variable across time or treatment(s).

Two-tailed test: A test measuring for either an increase or decrease in a test variable across a group(s).

A BRIEF INTRODUCTION TO THIS UNIT

The Wilcoxon signed rank test is usually referred to as the nonparametric version of the paired samples *t*-test. The signed rank test can be used to measure a significant change or shift in scores across time or after some intervention (pre-/posttest designs). The test is also appropriate for matched-pairs designs. As with the unit on the rank sum test, attention will be given at the end of this unit for conducting the signed rank test in SPSS. For the Wilcoxon signed rank test, SPSS provides an asymptotic (approximated) *p* value.

PROCEDURES FOR CONDUCTING A TWO-TAILED TEST FOR SMALLER SAMPLE SIZES

The Wilcoxon signed rank test is appropriate when the researcher wishes to measure for a significant change (*C*) from pretest to posttest (or pretreatment to posttreatment) when the alternative hypothesis is either directional (expected increase or decrease in ranks) or nondirectional, that the scores or ranks will change significantly either positively or negatively. In this unit, the null hypothesis will be represented by H_0: $C = 0$ and the alternative hypothesis by $C \neq 0$.

The Research Problem Testing H_0: $C = 0$

The US Environmental Protection Agency (EPA) keeps track of tier 1 criteria air pollutants (CAPS) from a variety of sources, including wildfires, prescribed fires, metals processing, solvent utilization, and highway and off-highway emissions. Consider a random sample of pollutants from two sources in California for the years 2019 and 2020 in table 4.1 ($n = 12^2$). Test the null hypothesis that no significant change in pollutant discharge occurred from 2019 to 2020. Conduct a two-tailed test with $\alpha = .055$.

Determine whether the Data Are Normally Distributed or Nonparametric in Their Shape

It is common for researchers, prior to conducting statistical analyses, to first check data sets for normality. Unlike the summated scores on pizzeria ranks, which were clearly ordinal in nature, the pollutant discharge amounts in this exercise are ratio (scale) data. While we practically always use nonparametric

---

2　Testing with $n = 12$ does not lend to a powerful test; $n = 12$ was chosen for demonstration purposes only.

tests for ordinal data, ratio data can be analyzed with either parametric or nonparametric tests; the decision is usually based on the results of a normality test. SPSS provides the output of two normality tests: the Kolmogorov–Smirnov test and the Shapiro–Wilk test. Rule of thumb is to consider the results of the Shapiro–Wilk test if $N \leq 50$ and to consider the results of the Kolmogorov–Smirnov test if $N > 50$. Some researchers will examine the shape of the histogram and the observed values as they appear on the Q-Q plot (also provided by SPSS) to support their decision on the normality of the variable in the population. If our data on CAPS are considered normally distributed, a paired samples t-test will suffice in testing the null hypothesis. If the Shapiro–Wilk test is significant ($p \leq .05$) and results of the histogram and Q-Q plot support this assumption (omitted in proof), the appropriate test is the Wilcoxon signed rank test. Both the Kolmogorov–Smirnov and the Shapiro–Wilk test measure the likelihood of the sample data coming from a normally distributed population.

With the data for both 2019 and 2020 entered as separate variables in SPSS, use ANALYZE > DESCRIPTIVE STATISTICS > EXPLORE. Move the data for both years into the Dependent List box, and in Plots, select Normality plots with tests and Histogram.

Results of both Shapiro–Wilk tests are significant, with $p < .001$ and $p < .001$, respectively. The conclusion is that the data are not normally distributed; the Wilcoxon signed rank test is the appropriate test to use here.

Step 1: Verify That Assumptions for the Test Are Met

Assumption 1: Each pair of scores or ranks must represent a random sample from a population and be independent of the other pairs in the sample.

Assumption 2: The distribution of the difference scores (D) are continuous with only few tied differences.

Step 2: State the Hypotheses

The null hypothesis is that there is no significant difference in satisfaction levels:

$$H_0: C = 0$$

The alternative hypothesis is that satisfaction levels changed:

$$H_1: C \neq 0$$

Step 3: Set the Critical Value for Rejection of the Null Hypothesis

In examining table C, when $n = 12$, with $\alpha = .055$, x (C_{cv}) = 60. As a two-tailed test, we need to measure for significant change in either direction.

Upper tail: Reject the null hypothesis if $T^+ \geq x$ (C_{cv}).
Upper tail: Reject the null hypothesis if $T^+ \geq 60$.
Lower tail: Reject the null hypothesis if $T^+ \leq n$ ($n + 1$) / 2 – x (C_{cv}).
Lower tail: Reject the null hypothesis if $T^+ \leq 12$ (13) / 2 – 60.
Lower tail: Reject the null hypothesis if $T^+ \leq 18$.

Step 4: Calculate T^+

The CAP amounts are provided for the years 2019 and 2020. Difference scores are symbolized by (D) and the absolute value of the difference scores shown as D. Ranks are based on absolute values, with the lowest difference value ranked 1. Positive change is coded 1 and negative change is coded 0. No ties occurred.

TABLE 4.1 **2019 and 2020 CAPS from CA Highway and Off-Highway Emissions Source**

	Pollutant	2019	2020	(D)	D	Rank	Code	Positive Ranks
Hwy	CO	633.77*	540.11	−93.66	93.66	11	0	
Hwy	Black Carbon	1.50	2.60	1.1	1.1	8	1	**8**
Hwy	NH3	10.75	10.01	−.74	.74	6	0	
Hwy	NOX	170.20	147.31	−22.89	22.89	10	0	
Hwy	Organic Carbon	2.56	2.91	.35	.35	4	1	**4**
Hwy	PM10-PRI	21.75	20.73	−1.02	1.02	7	0	
Hwy	PM25-PRI	10.12	9.49	−.63	.63	5	0	
Hwy	SO2	1.52	1.44	−.08	.08	2	0	
Hwy	VOC	79.71	69.25	−10.46	10.46	9	0	
Off-Hwy	CO	677.28	936.10	258.82	258.82	12	1	**12**
Off-Hwy	Black Carbon	3.70	3.45	−.25	.25	3	0	
Off-Hwy	NH3	.10	.17	.07	.07	1	1	**1**
			Sum of the positive ranks (T^+) =					**25**

*All emission values are in 1,000/tons.
CO = carbon monoxide; NH3 = ammonia; NOX = nitrogen dioxide; PM-10 PRI = particulate matter ≤ 10 mm; PM-25 PRI = particulate matter ≤ 2.5 mm; SO2 = sulfur dioxide; VOC = volatile organic compounds

Self-Check for Positive Ranks (T^+)

Of the 12 emissions evaluated, four showed an increase in amount. Summing the positive ranks, we calculate a T^+ of 25.

Step 5: Interpret the Results

Let us first review the criteria for rejecting the null hypotheses:

Upper tail: Reject the null hypothesis if $T^+ \geq 60$.
Reject the null hypothesis if $T^+ \leq 18$.

Since 25 is not greater than 60 and since 25 is not less than 18, we fail to reject the null hypotheses. We lack evidence for any significant change in CAPS from 2019 to 2020.

PROCEDURES FOR CONDUCTING A TWO-TAILED TEST FOR LARGER SAMPLE SIZES

While the table for upper-tail probabilities for the null distribution of the Wilcoxon signed rank T^+ statistic accommodates up to $n = 50$, sample sizes larger than 50 can be analyzed with a standardized formula that produces an approximate z value. We will use $T^+ = 25$ from the CAPS exercise ($n = 12$) example but recalculate using the formula for larger sample sizes. The formula is like that of the Wilcoxon W statistic and several other hypotheses tests in that we subtract the expected mean (when the null is true) from the observed mean and then adjust (divide) by the expected variance (when the null is true).

Calculation alert: The square root sign in this formula is designated by superscript .5 as seen outside and to the upper-right of the brackets:

$$z = T^+ - MT_0 / V_0$$

where $MT_0 = n\,(n + 1) / 4$
where $V_0 = [n\,(n + 1)\,(2n + 1) / 24]^{.5}$

$$MT^0 = 12\,(13) / 4 = 39$$
$$V_0 = [12\,(13)\,(25) / 24]^{.5} = 12.75$$

$$z = 25 - 39 / 12.75$$
$$z = -1.098$$

Since -1.098 is not to the left of -1.960, we fail to reject the null hypothesis. In examining table A, we see that when $z = -1.098$, $p = .1379$ for a one-tailed test. For a two-tailed test, $p = .2758\ (.1379 \times 2)$.

CALCULATING A REQUIRED SAMPLE SIZE

Let us return to G*Power for determining the sample size needed for a signed rank test given the researcher's criteria. Again, let us assume that the researcher sets alpha at .05, sets the power of the test $(1 - \beta)$ at .80, plans to conduct a one-tailed test, and desires to find a moderate effect size of .05.[3] Since this is a paired-samples test, there is only one group, so no group ratio needs to be calculated. The test family is *t-tests*, and the statistical test is *Means: Wilcoxon signed-rank test (matched pairs)*. According to the

---

3   A larger sample size is required to find a smaller effect size. Students can change the desired effect size from .50 to .80 (as used in unit 3) to see the difference.

calculations of G*Power, at least 28 subjects are needed to reach the actual power of .8083058 for this signed rank test. With that said, the researcher should strive to include as many subjects as possible, for power is increased when sample size is increased.

CONDUCTING A SIGNED RANK TEST IN SPSS

The first step in conducting a Wilcoxon signed rank test in SPSS is to define the variables. For demonstration purposes, we will use the same data (table 4.1) used for the CAPS study. After opening a new data field in SPSS, click on the Variable View tab in the lower-left corner of the field. Create two variables for the signed rank test. Since SPSS does not allow for numbers in the name column, we can enter *nineteen* and *twenty*. We need two spaces for decimals. Despite the CAPS being nonparametric in nature, they still are scale (ratio) data. Set accordingly.

TABLE 4.2   Variable View in SPSS Data Field

	Name	Type	Width	Decimals	Label	Values	Missing	Columns	Align	Measure	Role
	\textbf{Untitled 1 [Data Set0] – IBM SPSS Statistics Data Editor}										
1	nineteen	numerical	8	2	2019 CAPS	none	none	8	right	scale	input
2	twenty	numerical	8	2	2020 CAPS	none	none	8	right	scale	input
3											

Source: Generated with SPSS software. Copyright © 2023 by IBM Corporation. Reprinted with permission.

After clicking on the Data View tab, we can enter the data as shown in table 4.3.

TABLE 4.3   Data View in SPSS Data Field

	nineteen	twenty
	\textbf{Untitled 1 [Data Set0] – IBM SPSS Statistics Data Editor}	
1	633.77	540.11
2	1.50	2.60
3	10.75	10.01
4	170.20	147.31
5	2.56	2.91
6	21.75	20.73
7	10.12	9.49
8	1.52	1.44

9	79.71	69.25
10	677.28	936.10
11	3.70	3.45
12	.10	.17
13		

Once the data are entered, the process is as follows: ANALYZE > NONPARAMETRIC TESTS > LEGACY DIALOGS > 2 RELATED SAMPLES. At this point, you are provided with the test field. Highlight *nineteen* and move it inside the box for Variable 1. Highlight *twenty* and move it inside the box for Variable 2. Make sure the default is set on Wilcoxon. Click OK.

TABLE 4.4   Modified Results Section of the Wilcoxon Signed Rank Test

	N	Mean Rank	Sum of Ranks
Negative Ranks	8	6.63	53.00
Positive Ranks	4	6.25	25.00
Ties	0		
Total	12		

Test Statistics	
2020–2019	
z	−1.098
asymp. Sig. (2 tailed)	.272

The p value in the SPSS output of .272 is similar to the approximate p value of .2758 calculated when using the formula for larger sample sizes. Both values are greater than .055, which was calculated using the smaller sample size. Note that the z values of the SPSS output and the formula for larger sample sizes are both 1.098.

CREATING AN APA RESULTS SECTION FOR THE WILCOXON SIGNED RANK TEST

We will again use the data from the SPSS results above to report our findings:

> A Wilcoxon signed rank test was conducted to evaluate the null hypothesis that no significant change occurred in CAPS between 2019 and 2020. The test was not significant, $z = -1.098$, $p = .272$.

THE ONE-SAMPLE WILCOXON SIGNED RANK TEST

There are times when a researcher wishes to compare recently calculated scores from a sample group to known, established scores of a population. This situation could arise as the result of some new procedure that gives reason to believe that the known, established score no longer holds true. For example, let us assume that, for the past five years, the median argumentative essay score at an urban middle school has been 3.0, on an ordinal scale of 1 through 5 (recognizing that some other standardized tests are on an interval scale). After a substantial change in the method of teaching, argumentative essay writing has been put into place at the school for eight months, and the Language Arts Department chairperson wishes to know if the school's median argumentative essay score has changed for her school based on this sample with a chosen confidence level (e.g. $\alpha = .05$). Once a sample of the new essay scores are entered into the SPSS Variable Field as ordinal data, the procedures are as follows: ANALYZE > NONPARAMETRIC > ONE SAMPLE. Click on the Fields tab and move the new essay scores variable into the Field Tests box (if SPSS v. 29 has not already done so). Then click on the Settings tab. Choose Customize and then select Compare median to hypothesized (Wilcoxon signed-rank test); in the Hypothesized median box, enter 3.0. Click RUN. SPSS provides the decision on whether to reject the null hypothesis, various test statistics, and a bar graph illustrating the positioning of both the hypothetical median (established score of 3.0) and the observed median (the sample of new argumentative scores).

CONCLUSION

Researchers can use the Wilcoxon signed rank test to determine if a significant change in some ordinal or nonparametric measure has occurred because of an intervention, phenomenon, or time itself. Such measures may include attitudes or levels of agreement, pain scores, various rankings, or continuous measures (interval or ratio) that fail to meet the assumptions of a normal distribution test. Tables are available for small sample sizes and a standardized formula is available for larger sample sizes. The Wilcoxon signed ranks test can be conducted as either a one-tailed or two-tailed test. It is often referred to as nonparametric version of the paired samples t-test.

Exercises

1. To test the effectiveness of a new therapy, a pain score inventory was administered pretreatment and posttreatment on eight ($n = 8$) randomly selected elderly patients all suffering from lower extremity joint pain. Use the two-tailed formula for smaller sample sizes and values from table C to test the null hypothesis that there was no significant change in self-reported pain. Test with $\alpha = .055$. Use the following data set:

Subject	Pre-Score	Post-Score	(D)	D	Rank	Code: 1 = Positive 0 = Negative	Positive Ranks
1	32	28					
2	54	52					
3	50	28					
4	46	40					
5	46	41					
6	51	52					
7	37	26					
8	38	41					
						Σ positive ranks (T^+) =	

2. The counselors at Camp Serene investigate the effectiveness of a two-week summer camp session on the self-esteem of 12 underprivileged inner-city teens. They administer a 25-item self-esteem inventory (Likert scaling) one week prior to the start of camp and at the end of the two-week session. Summated scores are provided, with higher scores indicating higher self-esteem. Use the two-tailed larger sample sizes (standardized) formula to test the null hypothesis of H_0: $C = 0$. Test with $\alpha = .05$. Remember to reduce the value of n for any difference scores of 0. Use the data set provided.

Subject	Pre-Score	Post-Score	(D)	D	Rank	Code: 1 = Positive 0 = Negative	Positive Ranks
1	150	200					
2	160	200					
3	160	150					
4	170	260					
5	200	225					
6	210	210					
7	210	224					
8	215	214					
9	215	227					
10	220	226					

						Code: 1 = Positive 0 = Negative	Positive Ranks
11	220	240					
12	250	248					
						Σ positive ranks $(T^+) =$	

3. Researchers investigate the effectiveness of hypnosis on the reduction of cigarette smoking among a randomly selected group of Canadians who for the past five years have smoked one to two packs of cigarettes daily. Pre-scores are the average number of cigarettes smoked daily based on the week prior to hypnotic treatment; post-scores are the average cigarettes smoked daily based on the two weeks following treatment. Assume that the data are non-normal. Using SPSS, test the null hypothesis that cigarette consumption did not change after hypnotic treatment. Set $\alpha = .05$. Write an APA results section as your answer. Use the data set provided.

Subject	Pre-Hypnosis	Post-Hypnosis	(D)	D	Rank	Code: 1 = Positive 0 = Negative	Positive Ranks
1	32	33					
2	28	26					
3	40	40					
4	33	34					
5	32	21					
6	26	18					
7	24	17					
8	19	14					
9	20	15					
						Σ positive ranks $(T^+) =$	

4. If the goal of the hypnotic treatment in the previous exercise was to reduce daily cigarette usage, how satisfied or dissatisfied should the researchers be? How might a different population of smokers under study produce different results? To what length should a researcher go in describing one's population when conducting research?

5. The dean of discipline at a public middle school wants to know if a modified version of the current behavioral contract reduces the number of nonviolent incidents for the fifty students currently under contract. The data set is a randomly chosen subset of 15 students from the 50 students on contract. Counts reflect the five months prior to the modification and the five months after the modification. Assume that the data are non-normal. Using SPSS, test the null hypothesis that the number of nonviolent incidents did not change after modification of the contract. Set $\alpha = .05$ and use the asymptotic p value for your decision, with an alternative hypothesis of H_1: $C \neq 0$. Write an APA results section as your answer. Use the data set provided.

Subject	Pre-Change	Post-Change	(D)	D	Rank	Code: 1 = Positive 0 = Negative	Positive Ranks
1	6	4					
2	11	12					
3	10	14					
4	8	8					
5	12	13					
6	9	4					
7	14	9					
8	8	3					
9	13	14					
10	13	6					
11	5	6					
12	10	10					
13	7	4					
14	8	1					
15	11	12					
						Σ positive ranks (T^+) =	

6. This exercise contains a set of fictitious data. The answers to this exercise are *not* included in the Answers to Unit Exercises section of this text. You may want to consult your instructor for assistance or for confirmation of your results. Use SPSS to conduct the test, as table C does not include $n = 15$.

Studies have shown that time spent outdoors (TSO) can improve one's health and well-being. A mental health scale was administered to 15 young, urban professionals a day prior to spending ten days in a remote forest hiking, canoeing, preparing outdoor meals, and sleeping outdoors. The scale was again administered to the subjects 48 hours after leaving the forest. Conduct a signed rank test to test the null hypothesis that no change in mental health occurred. Set alpha = .05. A higher score on the scale (ordinal data) is indicative of better mental health. Run the test again without the scores of subject 15, who had the lowest score pretreatment but the highest score posttreatment. Do the results change or stay the same?

Subject	Score before TSO	Score after TSO	Subject	Score before TSO	Score after TSO
1	33	48	9	19	26
2	18	28	10	20	29
3	20	20	11	28	27
4	24	30	12	28	35
5	22	29	13	26	31

6	31	30	14	16	20
7	16	24	15	14	49
8	21	29			

Omnibus Tests for Three or More Independent Samples

OBJECTIVES

- Understand and apply procedures for the Kruskal–Wallis H test.
- Understand the procedures for correcting for ties.
- Analyze and create a results section for the Kruskal–Wallis H test.
- Conduct a Kruskal–Wallis H test using SPSS.

KEY SYMBOLS

k: Number of independent groups

n_1: Number of cases or subjects in a group 1 (with continuous subscript numbering for all groups)

$s_1 = \ldots = s_k$: Scores are equal for all groups

$x\ (H_{cv})$: The value corresponding to a particular k, set of n and α when the null hypothesis is true

H_{obs}: The observed or calculated value of H for smaller and larger sample sizes

KEY TERMS

Degrees of freedom: The maximum number of logically independent values in a data sample.

Omnibus test: A significance test that tests several parameters at once.

Post hoc analysis: A test used to identify specific differences between groups if an omnibus test is significant.

A BRIEF INTRODUCTION TO THIS UNIT

The Kruskal–Wallis H test is often referred to as the nonparametric version of the one-way analysis of variance (one-way ANOVA). It is an omnibus test in that post hoc analyses are required to know exactly which groups are significantly different than the others when the omnibus test is significant. The Kruskal–Wallis H test results in SPSS produce a test statistic, degrees of freedom, and an asymptotic (two-tailed) p value all adjusted for ties. The significance values for the pairwise comparisons (post hoc analyses) are automatically adjusted by the Bonferroni correction for multiple tests in SPSS. In this unit, we will first rely on the asymptotic p value of the H test in determining whether to reject the null hypothesis. If the null hypothesis is rejected, we analyze the adjusted significance values to determine significant differences in groups or pairs (pairwise comparisons).

When there are five or more cases in a cell group, we can use the X^2 distribution to establish the critical value for rejecting the null hypothesis. Use of the chi-square distribution requires that the researcher knows how many degrees of freedom there are. For the Kruskal–Wallis H test, use $k - 1$. For example, if the researcher is testing for any differences in scores of three populations (using samples from three populations), then $3 - 1 = 2$ degrees of freedom. A quick look at the X^2 table (table E) shows that, with two degrees of freedom and $\alpha = .05$, the critical value is 5.991. The researcher would then reject the null hypothesis if the observed H value were equal to or exceeded 5.991.

PROCEDURES FOR CONDUCTING A KRUSKAL–WALLIS H TEST FOR SMALLER SAMPLE SIZES

The selected upper-tail probabilities for the null distribution of the Kruskal–Wallis H statistic found in this text accommodates $k = 3$ with all $n = 3$ through all $n = 7$. In this unit, the null hypothesis for the omnibus test is $H_0: s_1 = ... = s_k$. In other words, the scores for all groups are the same.

The Research Problem Testing $H_0: s_1 = ... = s_k$

A medical researcher wants to know if pain scores will be the same across three different treatment groups: drug A, drug B, and a combination of drug A and B. Summated pain scores (a lower value indicating less pain) are collected from twenty-one patients randomly chosen and randomly placed into one of three treatment groups, with seven patients in each group.

Step 1: Verify That the Assumptions Are Met

Assumption 1: The distribution of scores in both groups are similar in shape.[1]

Assumption 2: The scores are random samples from the respective populations and the scores are independent of one another.

---

1   Like the rank sum test, violation of this assumption means that rejection of the null is an indicator that the scores are significantly different but that the median scores are not necessarily different.

Step 2: State the Hypotheses

The null hypothesis is that pain scores are the same for all three groups:

$$H_0: s_1 = ... = s_k$$

The alternative hypothesis is that pain scores for some groups are different:

$$H_0: s_1 \neq ... \neq s_k$$

Step 3: Set The Critical Value for Rejection of the Null Hypothesis

In examining table D, it is clear that when $k = 3$ and $n_1 = 7$, $n_2 = 7$, and $n_3 = 7$ and that $\alpha = .0506$, $x(H_{cv}) = 5.766$. Reject the null hypothesis if $H_{obs} \geq 5.766$.

Step 4: Calculate the Observed Value of *H*

To calculate the observed value of *H*, we first need to rank the scores from lowest to highest across all three groups and then sum the ranks for each group. The rankings found in table 5.1 are in parentheses.

TABLE 5.1   Self-Reported Pain Scores for Three Treatment Groups

Drug A	Drug B	Drug A & B
22 (14)	29 (19)	30 (20)
17 (9)	28 (18)	26 (16)
16 (8)	27 (17)	25 (15)
15 (7)	20 (12)	21 (13)
14 (6)	19 (11)	18 (10)
13 (5)	11 (3)	31 (21)
12 (4)	10 (2)	09 (1)
$\Sigma R_1 = 53$	$\Sigma R_2 = 82$	$\Sigma R_3 = 96$

Self-Check for Rankings

When N = the total number of scores (in this case, 21), the self-check formula equals the sum of the sum of the ranks (in this case, 53 + 82 + 96 = 231). So,

$$.5N (N + 1) = 231$$
$$.5 (21)(22) = 231$$
$$231 = 231$$

It appears that we have accurately ranked and summed the data.

Step 4 Continued

Continuing to calculate the observed value of H (H_{obs}), we will use the following calculation:

$$H_{obs} = 12 / N(N + 1) [(\Sigma R_1)^2 / n_1 + (\Sigma R_2)^2 / n_2 + (\Sigma R_3)^2 / n_3] - 3 (N + 1)$$
$$H_{obs} = 12/462 [(53^2) / 7 + (82^2) / 7 + (96^2) / 7)] - 3(22)$$
$$H_{obs} = 3.667$$

Step 5: Interpret the Results

First, let us review the criterion for rejecting the null hypothesis:

$$\text{Reject the null hypothesis if } H_{obs} \geq 5.766.$$

Since 3.667 is not greater than 5.766, we fail to reject the null. We lack evidence that pain scores are different across treatments when $\alpha = .0506$.

PROCEDURES FOR CONDUCTING A KRUSKAL–WALLIS H TEST FOR LARGER SAMPLE SIZES

As mentioned in the brief introduction for this unit, when there are five or more samples for each population under scrutiny, the researcher can refer to the x^2 distribution for the critical value of rejecting the null. Consider the following data set, assumed to be non-normal *with ties*; the researcher investigated for differences in cherry tree growth in centimeters over a year period between four different soil types. Rank sums (ΣR) and average ranks () have been calculated. We will calculate the H value (H_{obs}):

$$H_{obs} = 12 / N(N + 1) [(\Sigma R_1)^2 / n_1 + (\Sigma R_2)^2 / n_2 + (\Sigma R_3)^2 / n_3] - 3 (N + 1)$$
$$H_{obs} = 12 / 420 [(90^2) / 5 + (46.5^2) / 5 + (38^2) / 5 + (35.5^2 / 5)] - 3(21)$$
$$H_{obs} = 10.909$$

TABLE 5.2   Yearly Growth of Cherry Trees in Centimeters for Four Soil Types

Type 1	Type 2	Type 3	Type 4
45 (20)	36 (15)	34 (13)	33 (12)
44 (19)	35 (14)	31 (10)	32 (11)
43 (18)	29 (8.5)	29 (8.5)	28 (7)
40 (17)	27 (6)	26 (5)	25 (4)
39 (16)	24 (3)	22 (1.5)	22 (1.5)
$\Sigma R_1 = 90$ $\bar{R}_1 = 18$	$\Sigma R_2 = 46.5$ $\bar{R}_2 = 9.3$	$\Sigma R_3 = 38$ $\bar{R}_3 = 7.6$	$\Sigma R_4 = 35.5$ $_4 = 7.1$
$H_{obs} = 10.909$			

We need to know the degrees of freedom and alpha a priori before referring to table E for the critical value for rejecting the null hypothesis. For degrees of freedom, we have $4 - 1 = 3$ $(k - 1)$. With $\alpha = .05$ and three degrees of freedom, the critical value is 7.815. Since 10.909 exceeds the critical value of 7.815, we can reject the null hypothesis for the omnibus test of the H statistic. At least one of the soil types leads to significantly different growth levels than another.

Correcting for Ties

The formula for correcting for ties is as follows: $1 - (\Sigma t3 - t) / N3 - N$, with N = the total number of subjects in the study.

Let us use table 5.3 to better illustrate the calculation of $\Sigma t^3 - t$.

TABLE 5.3

Values Tied	22	29
Number Observed	2	2
$t^3 - t$	$2^3 - 2 = 6$	$2^3 - 2 = 6$
$\Sigma t^3 - t$	6 + 6	12

The calculation for the adjustment value is now as follows: $1 - 12 / 20^3 - 20$.

$$1 - (11 / = 7980) = .9987.$$ To adjust for ties using .9987, divide $H_{obs} = 10.909$ by the adjustment: adjusted $H_{obs} = 10.909 / .9987 = 11.004$[2]

Pairwise Comparisons

When rejecting the null hypothesis of no differences in growth among the four soil types, we need a follow-up (post hoc) method of testing all pairwise comparisons. Since we have four soil types, we need to make six pairwise comparisons, all as follow-up tests:

Type 1–type 2
Type 1–type 3
Type 1–type 4
Type 2–type 3
Type 2–type 4
Type 3–type 4

Many methods of conducting pairwise comparisons exist. SPSS, as shown later in this unit, conducts Wilcoxon–Mann–Whitney rank sum tests with p values adjusted by the Bonferroni correction for multiple comparisons.

---

2   SPSS calculates the adjusted for ties H value as 11.111, the difference likely due to rounding issues.

CONDUCTING A KRUSKAL–WALLIS *H* TEST IN SPSS

The first step in conducting the Kruskal–Wallis *H* test in SPSS is to define the variables. For demonstration purposes, we will use the same data as found in table 5.2 (cherry tree growth example). After opening a new data field in SPSS, click on the Variable View tab in the lower-left corner of the field. Create two variables for the Kruskal–Wallis *H* test: *soil* and *growth* (see table 5.4).

TABLE 5.4   Variable View in SPSS Data Field

	Name	Type	Width	Decimals	Label	Values	Missing	Columns	Align	Measure	Role
				Untitled 1 [Data Set0] – IBM SPSS Statistics Data Editor							
1	*soil*	numerical	8	0	*soil*	1, Ty	none	8	right	*nominal*	input
2	*growth*	numerical	8	0	*growth*	none	none	8	right	*scale*	input
3											

Source: Generated with SPSS software. Copyright © 2023 by IBM Corporation. Reprinted with permission.

For *soil*, type in *soil* for Name, and the data Type is numerical[3]; no decimals are needed. In the Label column, the researcher can enter a lengthier description, such as *type of soil*. In the Values box, enter 1 and label it Type 1; enter the subsequent numbers of 2, 3, and 4 for the other three soil types. Click **OK**. Drag down the Measure box to make *soil* nominal, as we are naming the groups.

For Name, type in *growth*, then enter *yearly growth in centimeters* in the Label box.[4] We can adjust decimals to 0, as the data set does not contain values with decimals. No Values are needed, as growth is its own value. Drag down your choices for Measure and set it at *scale*, as growth in centimeters is scale data.[5]

Now click the *Data View* tab in the lower-left corner of the data field. This is where we enter the data for the growth in centimeters example (table 5.5).

TABLE 5.5   Data View in SPSS Data Field

	soil	growth
	Untitled 1 [Data Set0] – IBM SPSS Statistics Data Editor	
1	1	45
2	1	44

---

3   When importing data from an existing Excel spreadsheet, the data type is sometimes identified as *string*. In such a case, simply drag down the options and set to *numerical*.
4   The Label box can be expanded quite a bit to allow for a lengthier description of the variable. It is the Label description that will appear in most results sections.
5   Growth in centimeters is considered scale data because it meets the criteria of both interval and ratio data: having numbers as values, having equal distances between consecutive numbers, and having a meaningful zero. Researchers analyzing either interval or ratio data in SPSS should select scale as the measure.

		Untitled 1 [Data Set0] – IBM SPSS Statistics Data Editor	
3	1	43	
4	1	40	
5	1	39	
6	2	36	
7	2	35	
8	2	29	
9	2	27	
10	2	24	
11	3	34	
12	3	31	
13	3	29	
14	3	26	
15	3	22	
16	4	33	
17	4	32	
18	4	28	
19	4	25	
20	4	22	

Source: Generated with SPSS software. Copyright © 2023 by IBM Corporation. Reprinted with permission.

Once the data are entered, the drag-down process is as follows: ANALYZE > NONPARAMETRIC TESTS > INDEPENDENT SAMPLES. At this point, you are provided with three tabs, a conditional statement, and choice of objectives. The conditional statement reminds the researcher that, at this point, any tests conducted are based on the assumption that your data do not follow the normal distribution. Since we are more concerned about differences in growth distributions rather than a specific median,[6] leave the objective at its default: Automatically compare distributions across groups. Click the Fields tab. Move soil down to the Groups box; move growth over to the Test Fields box. Then click the Settings tab. Choose Customize tests, and then choose Kruskal–Wallis 1-way ANOVA (k samples). For Multiple comparisons, choose All pairwise. Now click **RUN.** The results are in table 5.6.

---

6   Remember that our null hypothesis is that scores are different across soil types. If the study designed called for a comparison of medians, we could have chosen to compare median scores.

TABLE 5.6   Modified Results Section of the Kruskal–Wallis One-Way ANOVA[7]

Hypothesis Test Summary				
	Null Hypothesis	Test	Sig.[a, b]	Decision
1	The distribution of growth in centimeters is the same across categories of soil type	Independent-Samples Kruskal–Wallis Test	.011	Reject the null hypothesis

a. The significance level is .050.

b. Asymptotic significance is displayed.

Independent-Samples Kruskal–Wallis Test Summary	
Total N	20
Test Statistic	11.111[a]
Degrees of Freedom	3
Asymptotic Sig. (2-sided test)	.011

a. The test statistic is adjusted for ties.

Pairwise Comparisons of Soil Type					
Sample 1-Sample 2	Test Statistic	Std. Error	Std. Test Statistic	Sig.	Adj. Sig.[a]
4-3	0500	3.739	.134	.849	1.000
4-2	2.200	3.739	.588	.556	1.000
4-1	10.900	3.739	2.915	.004	.021
3-2	1.700	3.739	.455	.649	1.000
3-1	10.400	3.739	2.782	.005	.032
2-1	8.700	3.739	2.327	.020	.120

a. Significance values have been adjusted by the Bonferroni correction for multiple tests.

Source: Generated with SPSS software. Copyright © 2023 by IBM Corporation. Reprinted with permission.

First, let's analyze the Hypothesis Test Summary. Unlike some SPSS output, the output for this procedure generates the null hypothesis and provides the correct decision on whether to reject the null based on $\alpha = .05$. We are also given the asymptotic *p* value. In the Independent-Samples Kruskal-Wallis Test Summary section, we see the calculated H value of 11.111, practically identical to our H value of 11.004, both the SPSS H value and ours having been adjusted for ties. Finally, we can determine which pairwise comparisons are significantly different in the Pairwise Comparison of soil type section. Note

---

7   SPSS refers to this test as both the Kruskal–Wallis 1-way ANOVA and Independent-Samples Kruskal-Wallis test as followed in the procedures. If the researcher chooses LEGACY DIALOGS > K INDEPENDENT SAMPLES, SPSS refers to the test as the Kruskal-Wallis *H*.

that in the far-right column, adjusted significance values have been provided for all six pairwise comparisons. Only two pairwise comparisons are significant at $\alpha = .05$: Type 1 soil produces significantly higher[8] cherry tree growth compared to either type 3 or type 4 soil. We have no evidence for any other significant differences among soil types.

CREATING AN APA RESULTS SECTION FOR THE KRUSKAL–WALLIS ONE-WAY ANOVA

For our results section, we will use the previous SPSS results:

> A Kruskal–Wallis H test was conducted to evaluate differences in cherry tree growth in centimeters between four soil types. The test, corrected for ties, was significant: H (N = 20) = 11.111, $p = .011$. Post hoc analyses were conducted to evaluate pairwise differences between the four soil types. Results indicated significant differences in growth between type 1 soil and type 3 soil (adj. $p = .032$) as well as type 1 soil and type 4 soil (adj. p value = .021). Higher growth was associated with type 1 soil.

OTHER OMNIBUS TESTS FOR THREE OR MORE INDEPENDENT SAMPLES

The Friedman Test

For researchers employing a repeated-measures design but with ordinal or non-normal data, the Friedman test[9] is an appropriate choice. It can be used when a researcher wishes to measure levels of agreement to some policy or social action (using a Likert scale) over time: before exposure to an educational seminar, after the seminar, and 12 weeks after the seminar. Like the Kruskal–Wallis H test, the Friedman test is an omnibus test requiring post hoc analyses if the test is significant. This is explained in detail when the procedures for conducting the test in SPSS are explained.

As another example, let us assume that a high school science department chairperson is interested in knowing which of three different methods of learning about climate change juniors and seniors believed was most effective in making students aware of its current, measurable negative effects. The three methods are film, historical fiction novel, and lecture with photos. Students are exposed to all three methods over the course of a school semester and then are asked to rate the effectiveness of each medium on a 1–10 scale, with a 10 being most effective. The test measures whether one medium is rated more effective than the others.

---

8   Since the output did not include descriptive statistics, we need to look again at the data set to see which soil type had higher levels of growth in centimeters. In this case, type 1 soil had the higher growth measures.
9   The Friedman test is sometimes referred to as the nonparametric version of the one-way repeated measures ANOVA, or Friedman's ANOVA.

If the science department chairperson is concerned that students will respond more favorably to the method of exposure closest to the time of the survey (sometimes called a *nuisance variable*), students can be randomly placed into three blocks, whereby order of exposure is different for each block. For a medium-sized high school, each block might contain 140 students. Such an outlay might look like what we see in table 5.7. The use of blocks controls for any effect order of exposure might have on student ratings—often referred to as the *carry-over effect*. This design is common in medical studies when the order of treatment, which is of no interest to the researcher, might influence the outcome. Yet in such a design, the assumption is that no interaction exists between blocks and treatments.

TABLE 5.7   Layout for a Friedman Test with Three Blocks

Blocks	Subjects	Scores on Film	Scores on Novel	Scores on Lecture
Block 1 Lecture as last exposure	n = 140			
Block 2 Novel as last exposure	n = 140			
Block 3 Film as last exposure	n = 139			

The procedure for the Friedman test in SPSS is as follows: ANALYZE > NONPARAMETRIC TESTS > LEGACY DIALOGS > K RELATED SAMPLES. All three variables (scores on the three different media) are then entered into the Test Variables box. In SPSS, the Friedman test is the default setting. Run the test. The result is an omnibus test, with its significance indicating only that at least two of the three methods differ in terms of effectiveness. For post hoc analyses, conduct Wilcoxon signed rank tests and adjust for multiple comparisons using Bonferroni's correction. For example, if you are testing with α = .05, divide .05 by the number of test variables—in this case, three (film, historical fiction novel, and lecture scores)—which equals .017. Therefore, only reject each of the signed rank pairwise comparisons (film-novel, film-lecture, and novel-lecture) if $\alpha \leq .017$.

The Jonckheere–Terpstra Test for Ordered Alternatives

Oftentimes, when the researcher administers a treatment, the alternative hypothesis is that of either increasing[10] or decreasing[11] treatment effects over time. With the Jonckheere–Terpstra test, we can establish the alternative hypothesis of either increasing or decreasing treatment effects before running the test. In this manner, the results of pairwise comparisons automatically inform us of whether the effects increased or decreased significantly over time. The choice of the Jonckheere–Terpstra test is precisely

---

10   Smallest to largest, in SPSS command terms.
11   Largest to smallest, in SPSS command terms.

below the Kruskal–Wallis one-way ANOVA test in SPSS and is conducted in similar fashion. For an alternative hypothesis of increasing effects, drag down *smallest to largest*. For an alternative hypothesis of decreasing effects, drag down *largest to smallest*.

CONCLUSION

The Kruskal–Wallis H test is an efficient method of determining if scores are different between three or more groups when the data are ordinal, ranked, or non-normal. Post hoc analyses for pairwise comparisons can be conducted by hand using various methods of calculation, or, if using SPSS, automatically conducted if the null hypothesis is rejected. For repeated measures designs, both the Friedman test and Jonckheere–Terpstra test are available in SPSS with output that is rather easily interpreted.

Exercises

1. The research and development division of a large food processing company wants to know if the presence of microorganisms in a popular cheese product varies based on three levels of preservatives used in the processing of the cheese. The data represent the number of days when microorganisms are first present in the product. Rank the data and calculate rank sums and mean ranks for all three groups. Use the procedures for smaller sample sizes, setting $\alpha = .0509$ to test the null hypothesis H_0: $s_1 = \ldots = s_k$. Use the data set provided. Use the critical H values in table D. State whether differences exist between at least one pair of preservative levels.

Low	Rank	Moderate	Rank	High	Rank
31		25		34	
30		26		32	
28		20		27	
33		23		29	
18		17		19	
ΣR_1		ΣR_2		ΣR_3	
$\bar{R}_1$		$\bar{R}_2$		$\bar{R}_3$	

2. A psychologist at a large, public university compared self-reported levels of anxiety between freshmen taking end-of-semester exams under three exam-taking conditions. The data are summated anxiety scores taken from a ten-item inventory, with higher scores indicating higher levels of anxiety. Rank the data and calculate rank sums and mean ranks for all three groups using the data provided. Use the procedures for larger sample sizes to test the null hypothesis H_0: $s_1 = \ldots = s_k$. Refer to the X^2 table

for the critical value with $\alpha = .05$ and two degrees of freedom ($df = 2$). State if there are differences between at least one pair of conditions.

Large Auditorium	Rank	Computer Lab	Rank	Classroom	Rank
46		26		27	
44		24		25	
40		22		21	
39		20		19	
38		18		17	
37		16		15	
ΣR_1		ΣR_2		ΣR_3	
$\bar{R}_1$		$\bar{R}_2$		$\bar{R}_3$	

3. A superintendent of a large school district investigated the cognitive-emotional-behavioral support levels provided to students from 45 randomly chosen teachers from three types of schools: elementary, middle, and high school. A 24-item teacher evaluation tool using Likert scaling was developed and used during the evaluations. A higher score indicates a higher level of support. Use SPSS to test the null hypothesis that support levels provided are the same for all three types of schools. Examine post hoc analyses as appropriate. Use $\alpha = .05$. Use the data set provided. Report the findings in an APA-formatted results section.

Elementary School Teachers	C-E-B Score	Middle School Teachers	C-E-B Score	High School Teachers	C-E-B Score
1	41	1	39	1	30
2	40	2	44	2	29
3	36	3	37	3	29
4	31	4	37	4	25
5	28	5	29	5	24
6	24	6	29	6	21
7	24	7	21	7	20
8	23	8	21	8	20
9	21	9	19	9	17
10	19	10	17	10	16
11	18	11	17	11	14
12	17	12	16	12	11

Elementary School Teachers	C-E-B Score	Middle School Teachers	C-E-B Score	High School Teachers	C-E-B Score
13	16	13	9	13	11
14	8	14	10	14	9
		15	11	15	9
				16	9

4. This exercise is based on data from the US Department of Transportation. The answers to this exercise are *not* included in the Answers to Unit Exercises section of this text. You may want to consult your instructor for assistance or for confirmation of your results. You can conduct a Kruskal–Wallis *H* test by hand using table D with $\alpha = .0506$ or by using SPSS with $\alpha = .05$.

Following are hours of delays from 2015 to 2021, either caused by Amtrak,[12] a host rail company,[13] or other.[14] Using these seven years as a random sample of Amtrak performance, test the null hypothesis that hours of delays are the same for all three causes for all years. Set $\alpha = .0506$ or $.05$ depending on method of analysis.

Year	Amtrak	Host	Other
2015	31,582	57,701	12,774
2016	26,339	48,555	15,087
2017	27,451	53,332	14,250
2018	26,967	55,217	14,194
2019	30,589	53,702	12,670
2020	15,690	41,868	15,707
2021	12,080	34,397	12,597

5. This exercise contains a fictitious data set. The answers to this exercise are not included in the Answers to Unit Exercises section of this text.

Attendance at county fairs can vary, with higher attendance usually during the weekends when fairgoers are free from work and school. Conduct a Friedman test in SPSS to test the null hypothesis that no change in overall county fair attendance occurred over a three-day period using a fictitious sample of fair attendance at seven fictitious counties in the US Midwest. Set alpha = .05. Conduct post-hoc analyses as appropriate.

---

12 Amtrak-owned tracks.
13 Track not attributed to Amtrak.
14 Customs, immigration, law enforcement, weather, or waiting for scheduled departure time.

Prior to conducting the Friedman test, conduct Shapiro–Wilk tests for each day's attendance and identify the two data sets that are non-normal by reporting all Shapiro–Wilk p values. Also comment on how the performance of a popular, local country singer at the Sycamore County Fair on Thursday night may have affected the results of any post hoc analyses.

County Fair Site	Thursday's Attendance	Friday's Attendance	Saturday's Attendance
Lyons County, KY	2,560	2,959	3,060
Bass County, OH	2,561	2,860	3,061
Free County, MI	1,840	2,240	2,341
Corn County, IL	1,862	2,262	2,362
Sycamore County, IL	4,600	3,700	3,811
Raven County, OH	1,888	3,555	2,388
Clarke County, IN	1,834	2,964	2,334

Correlational Measures

OBJECTIVES

- Understand and apply procedures of Kendall's tau.
- Understand and apply procedures of Spearman's rho.
- Analyze and create results sections for the two analyses.
- Conduct Kendall's tau and Spearman's rho in SPSS.

KEY SYMBOLS

τ: Kendall's tau

ρ: Spearman's rho

KEY TERMS

Concordant pairs: The number of observed ranks below a rank that is larger than that rank.

Discordant pairs: The number of observed ranks below a rank that is smaller than that rank.

Range of a correlation coefficient: –1.0 to 1.0.

A BRIEF INTRODUCTION TO THIS UNIT

There are many times when the catalyst for an in-depth study begins with the discovery of a significant correlation between two variables. While the discovery of a significant correlation does not necessarily mean that one variable has a positive or negative effect (causation) on the other, it *could* lead to the discovery of a causal relationship. That is why, in part, many researchers wish to test two variables to determine if a significant correlation (association) exists, because a correlation *can* sometimes also be causal.

The two methods of calculating the strength of a relationship between two ordinal or ranked variables in this unit differ in some respects.[1] Perhaps the most often-cited difference between Kendall's tau and Spearman's rho is that Spearman's rho will lead to a higher coefficient value. This is not always true, as one or two large deviations between a few pairs of variables can lead to a smaller coefficient when calculating Spearman's rho. Another important difference is that, because of the way in which Kendall's tau is calculated, we cannot justifiably square the coefficient value to determine the shared variance; with both Pearson's coefficient of correlation[2] and Spearman's rho, we can square the coefficient to determine how much of the variance is shared by the two variables.[3] For example, a significant, positive relationship between high school students' SAT reading scores and first-semester college GPA of $r = .69$ would produce a shared variance (coefficient of determination) of $r^2 = .476$. Perhaps the most important difference between the two procedures discussed in this unit is in the calculation of the coefficient. With Kendall's tau, we evaluate the number of concordant and discordant pairs of variables; with Spearman's rho, we evaluate the sum of the deviations squared.

PROCEDURES FOR CALCULATING A KENDALL'S TAU COEFFICIENT OF CORRELATION

Before we begin calculating the coefficient for Kendall's tau, let us better understand what concordant and discordant pairs are. Concordant pairs are the number of observed ranks below a rank that are *larger than that rank*. Discordant pairs are the number of observed ranks below a rank that are *smaller than that rank*. Let's refer to table 6.1 for clarification, where we have five subjects, each with a score A and score B and no ties for either score. Note also that score A is in ascending order[4] and that all scores are ranked, with the lowest score receiving a rank of 1 and the highest rank receiving a rank of 5.

---

1   There is much research published on the similarities, differences, advantages, and disadvantages of Kendall's tau versus Spearman's rho. Not all will be discussed in this unit.
2   Pearson's product-moment correlation coefficient is the measure of strength between two variables when the variables are either interval or scale data and have a linear relationship.
3   Note that when we square the coefficient of Spearman's rho, we are measuring the shared variance of the ranks and not the original data.
4   Use this standard method of listing the scores or ranks of the first column in ascending order, making sure that the pairs remain the same.

TABLE 6.1 Illustration of Concordant and Discordant Pairs

Subject	Score A	Rank	Score B	Rank	Concordant	Discordant
1	20	1	40	3	2	2
2	22	2	42	4	1	2
3	24	3	43	5	0	2
4	26	4	36	2	0	1
5	28	5	32	1		
					C = 3	D = 7

Let's start with the rank of 3 for subject 1. When looking down the column below the 3, we count two ranks larger than 3 (the 4 and 5) and two ranks smaller than 3 (the 1 and 2). Those values are then entered into the two far left columns, respectively. Looking down the column below the 4, we count one rank larger than 4 (the 5) and two ranks smaller than 4 (the 2 and 1). This process is continued until the next to last rank, as *the 1 has no values* below it.

To calculate the tau coefficient, use the following formula:

tau = C – D / C + D

tau = 3 – 7 / 3 + 7

tau = 4 / 10

tau = –.40

But just because we can calculate a coefficient of –.40 in this situation does not mean that the value is significant. In other words, we need to conduct a hypothesis test to determine if the coefficient is significant at, say, .05. Fortunately, a standardized formula exists to determine if the tau coefficient is significant, with either a one-tailed or two-tailed test. Calculation alert: The square root sign in this formula is designated by superscript .5 as seen outside and to the upper-right of the brackets. The formula is as follows:

$z = (3)\ (tau)\ [n(n-1)]^{.5}\ /\ [2(2n+5)]^{.5}$

$z = (3)\ (-.40)\ [20]^{.5}\ /\ [30]^{.5}$

$z = (-1.2)\ (4.47)\ /\ 5.47$

$z = -.980$

which is nonsignificant at either –1.645 (one-tailed test) or –1.96 (two-tailed test).

For more practice, let us consider a situation in which a fitness coach at a community college investigated the relationship between upper-leg strength and ranked outcomes in a 200-meter dash event. Consider the upper-leg–strength values to be ordinal and a ranked outcome of 1 representing having won the 200-meter race[5] and a ranking of 8 having come in last among eight participants. The lowest leg strength is given the rank of 1; the highest a rank of 8.

_____

5   We are following the standard procedure of listing the first column of score/ranks in ascending order.

TABLE 6.2   Relationship between Placement in 200-Meter Dash and Leg Strength

Athlete	Rank in Race	Leg Strength	Strength Rank	C	D
1	1	66	8	0	7
2	2	61	6	1	5
3	3	58	5	1	4
4	4	56	4	1	3
5	5	62	7	0	3
6	6	54	3	0	2
7	7	50	2	0	1
8	8	46	1		
				C = 3	D = 25

Self-Check for Concordant and Discordant Pairs

The 4 has one concordant pair: the 7 below it (*larger* than the 4). The 4 has three discordant pairs below it: the 3, 2, and 1 (all *smaller* than the 4).

Procedures Continued

We can now calculate the coefficient:

$$tau = C - D / C + D$$
$$tau = 3 - 25 / 3 + 25$$
$$tau = -.786^6$$

Significance Test of Kendall's Tau

We now need to know the significance of the association. If the association lacks significance, it is rather meaningless to a researcher who wishes to generalize from the results.[7] Let us again calculate the z value, with $\alpha = .05$ and a two-tailed test, as we don't know whether upper-leg strength is related to doing well or doing poorly (two-sided test) in the 200-meter race. So, reject the null hypothesis if $z \leq 1.96$ or if $z \geq -1.96$.[8]

Calculation alert: The square root sign in this formula is designated by superscript .5, as seen outside and to the upper right of the brackets.

$$z = (3) (tau) [n(n - 1)]^{.5} / [2(2n + 5)]^{.5}$$
$$z = (3) (-.786) [56]^{.5} / [42]^{.5}$$

---

6   Calculating Spearman's rho, the coefficient *is* larger (-.857), as many experts would predict. But then this data set lacks pairs with large deviations.
7   When a study design is sound (e.g., proper selection and placement procedures and the appropriate test), researchers can often feel confident about using the results as a basis for making larger decisions.
8   That is to say, if z is a larger negative number—or to the left of -1.96 in the distribution.

$$z = (-2.36)(7.48) / 6.48$$
$$z = -2.724$$

Since –2.724 is to the left of –1.96 (a larger negative number), we reject the null hypothesis. We have evidence that a significant relationship exists between the two variables.

Interpret the Results

We have evidence that a significant relationship exists between placement in a 200-meter race and upper-leg strength. But what type of relationship? Since the coefficient is negative, we *must* say that, as one variable increases, the other decreases. In this case, as upper-leg strength increases, placement, or rank in the 200-meter race, decreases—which is a good thing for competitors with higher levels of upper-leg strength! Remember, the lower the race ranking, the better the outcome, as we assigned the first-place winner with a 1.

PROCEDURES FOR CALCULATING A SPEARMAN'S RHO COEFFICIENT OF CORRELATION

When calculating Spearman's rho (ρ), we do not rely on values of concordant and discordant pairs. Rather, we use the squared deviations of the pairs after ranking the data. Let us consider a situation in which a researcher investigated the relationship between class rank and hours spent in the library for the semester (consider hours to be non-normal; table 6.3). The top ten students in the class are evaluated.

TABLE 6.3   Class Rank and Hours Spent in the Library

Subject	Class Rank (X)	X Rank	Hours (Y)	Y Rank	D	D^2
1	1	1	400	10	–9	81
2	2	2	366	7	–5	25
3	3	3	200	1	2	4
4	4	4	340	6	–2	4
5	5	5	222	2	3	9
6	6	6	380	8	–2	4
7	7	7	390	9	–2	4
8	8	8	335	5	3	9
9	9	9	321	4	5	25
10	10	10	300	3	7	49
						$\Sigma d^2 = 214$

To calculate Spearman's rho coefficient, use the following formula:

$$rho = 1 - [6 \, (\Sigma d^2) \, / \, n \, (n^2 - 1)]$$
$$rho = 1 - [6 \, (214) \, / \, 10 \, (99)]$$
$$rho = 1 - [1284 \, / \, 990]$$
$$rho = 1 - 1.296$$
$$rho = -.296 \text{[9]}$$

Significance Test of Spearman's Rho

Is the value of −.296 statistically significant? As we did for Kendall's tau, we need to calculate the significance of this relationship. Here, we will apply Ramsey's (1989) formula and his use of the t-distribution to determine whether the calculated correlation between class rank and hours is significant. We will use the symbol t to represent the calculated value and refer to the t-distribution for the critical value when $\alpha = .05$ with a two-tailed test with n − 2 degrees of freedom. Note the square root signs of superscript .5:

$$t = \frac{(rho) \, [n - 2]^{.5}}{[1 - [rho^2]^{.5}}$$
$$t = \frac{(-.296) \, [10 - 2]^{.5}}{[1 - .0876]^{.5}}$$
$$t = \frac{(-.296) \, (2.828)}{[.9124]^{.5}}$$
$$t = \frac{-.837}{.955}$$
$$t = -.876$$

With the t-distribution (not included in this handbook but published in many textbooks and found on free website sources), when $\alpha = .05$ with a two-tailed test with 8 degrees of freedom (n − 2), the critical value is 2.306. The calculated value of .876 (consider the absolute value here) does not exceed the critical value of 2.306. The null stands. We have no evidence of a significant correlation between class rank and library hours based on our sample. Be advised that this formula works best when n ≥ 10.

This decision to conduct a two-tailed test was arbitrary. If the researcher had reason to believe that library hours are likely to increase class rank through prior research or knowledge of the learning characteristics at this mock university, a one-tailed test would be warranted. Such a decision must be made a priori. Researchers should not "shop around" for a desired p value by considering the results of a one-tailed test if the two-tailed test does not demonstrate significance. Such a tactic would undermine the integrity of the research field.

---

9   We will rely on SPSS for the significance test.

CONDUCTING KENDALL'S TAU AND SPEARMAN'S RHO IN SPSS

The first step in conducting either Kendall's tau or Spearman's rho in SPSS is to define the two variables. For demonstration purposes, we will use the data from table 6.2, with the two variables of race placement and upper-leg strength. After opening a new data field in SPSS, click on the Variable View tab in the lower-left corner of the field. Create two variables: *rank* and *strength* (table 6.4).

TABLE 6.4   Variable View in SPSS Data Field

	Name	Type	Width	Decimals	Label	Values	Missing	Columns	Align	Measure	Role
					Untitled 1 [Data Set0] – IBM SPSS Statistics Data Editor						
1	*rank*	numerical	8	0	*class r*	none	none	8	right	*ordinal*	input
2	*stren-gth*	numerical	8	0	*upper*	none	none	8	right	*scale*	input
3											

Source: Generated with SPSS software. Copyright © 2023 by IBM Corporation. Reprinted with permission.

Again, no need for decimals. In Label, we can expand on the variable names, so enter *class rank* and *upper-leg strength*. The ranks will be their own values, so use the Values default of none. Both *variables* are *ordinal* in nature.

Now click the Data View tab. This is where we enter the data for either Kendall's tau or Spearman's rho (table 6.5).

TABLE 6.5   Data View in SPSS Data Field

	rank	strength
		Untitled 1 [Data Set0] – IBM SPSS Statistics Data Editor
1	1	66
2	2	61
3	3	58
4	4	56
5	5	62
6	6	54
7	7	50
8	8	46

Source: Generated with SPSS software. Copyright © 2023 by IBM Corporation. Reprinted with permission.

Once the data are entered, the drag-down process is as follows: ANALYZE > CORRELATE > BIVARIATE. Move both variables into the Variables box to the right. Choose Kendall's tau-b,[10] as the default is set on Pearson. In this case, choose the default of a *two-tailed* Test of Significance, as we don't have an alternative hypothesis about the effects of upper-leg strength on performance in the 200-meter race. The results are in table 6.6.

TABLE 6.6 **Modified Results Section of Kendall's Tau-b**

Correlations			200-meter placement	Upper-leg stren-gth
Kendall's tau-b	200-meter placement	Coefficient Correlation	1.000	–.786**
		Sig. (2-tailed)	.	.006
		N	8	8
	Upper-leg stren-gth	Coefficient Correlation	–.786**	1.000
		Sig. (2-tailed)	.006	.
		N	8	8

** Correlation is significant at the 0.01 level (2-tailed)

Source: Generated with SPSS software. Copyright © 2023 by IBM Corporation. Reprinted with permission.

These are the same results that we calculated by hand in the earlier portion of this unit. Based on the output of SPSS, we consider the relationship to be significant at $\alpha = .006$ and deem the relationship to be a high, negative correlation of –.786.

The results of a Spearman's rho conducted in SPSS will look quite similar, except for the larger coefficient and slightly different significance level (table 6.7).

TABLE 6.7 **Modified Results Section of Spearman's Rho**

Correlations			200-meter placement	upper-leg stren-gth
Spearman's rho	200-meter placement	Coefficient Correlation	1.000	–.857**
		Sig. (2-tailed)	.	.007
		N	8	8
	upper-leg stren-gth	Coefficient Correlation	–.857**	1.000

---

10   Kendall's tau-b is used in SPSS because it corrects for ties if applicable. We used the formula for tau-a here.

	Sig. (2-tailed)	.007	.
	N	8	8

** Correlation is significant at the 0.01 level (2-tailed)

CREATING AN APA RESULTS SECTION FOR KENDALL'S TAU-B AND SPEARMAN'S RHO

We will use the results from table 6.6s and 6.7 to report the findings APA style:

> A Kendall's tau-b test was conducted to determine whether a significant relationship exists between 200-meter placement and upper-leg strength. The test was significant, $p = .006$, $\tau = -.786$. A high, negative correlation exists, with race placement decreasing (better performance) as upper-leg strength increases.

> A Spearman's rho test was conducted to determine whether a significant relationship exists between 200-meter placement and upper-leg strength. The test was significant, $p = .007$, $\rho = -.857$. A high, negative correlation exists, with race placement decreasing (better performance) as upper-leg strength increases.

CONCLUSION

Researchers have several options for conducting correlational analyses. While not covered in this handbook but available in SPSS, Pearson's correlation coefficient is the appropriate analysis when the relationship between the two variables is linear, and the quantitative data are normally distributed. When the data are ordinal or non-normal, either Kendall's tau or Spearman's rho is appropriate to analyze. When conducting Kendall's tau in SPSS, the user has the option of conducting a one-tailed or two-tailed test; the significance of either test is reported along with the correlation coefficient. When conducting Spearman's rho in SPSS, the same options apply, and the output is the same.

Exercises

1. Determine whether a significant relationship exists between the age of senior homeowners and level of agreement to conserve water by conducting a Kendall's tau-b test using SPSS and the considering the two-tailed test results. Report the findings in an APA results section. Remember to use a homeowner's summated score as the second variable.

Following are ages and summated likelihood scores of 51 randomly selected senior homeowners to replace grass lawns with xeriscaping. Summated scores of ≥ 14 are indicative of the homeowner following through and replacing a grass lawn with xeriscaping.

Likert scaling (three-item survey):
1 = very unlikely, 2 = unlikely, 3 = somewhat likely, 4 = likely, 5 = very likely
    Item 1: I plan on soon taking steps to reduce my water consumption.
    Item 2: I can learn to appreciate the aesthetics of xeriscaping.
    Item 3: I can learn to enjoy working in a xeriscaping garden or spending time in a xeriscaping garden.

Subject	Age	Item1	Item2	Item3	Sumscore	Subject	Age	Item1	Item2	Item3	Sumscore
1	62	5	5	5	15	26	89	4	3	3	10
2	62	5	5	5	15	27	66	4	2	3	9
3	62	5	5	5	15	28	64	3	3	3	9
4	62	5	5	4	14	29	81	4	2	3	9
5	62	5	5	4	14	30	82	4	3	2	9
6	62	5	5	4	14	31	74	3	3	3	9
7	84	5	4	5	14	32	73	3	3	2	8
8	84	5	4	4	13	33	79	3	3	2	8
9	77	5	4	4	13	34	89	3	2	3	8
10	74	5	4	4	13	35	65	3	2	3	8
11	69	5	4	5	14	36	68	3	2	2	7
12	68	5	4	5	14	37	66	3	2	2	7
13	67	5	4	4	13	38	66	3	1	3	7
14	63	5	3	4	12	39	74	3	3	1	7
15	64	5	3	4	12	40	70	3	1	2	6
16	65	4	4	4	12	41	70	3	1	2	6
17	88	4	4	3	11	42	90	3	2	1	6
18	87	4	3	4	11	43	88	3	2	1	6
19	65	4	3	4	11	44	84	2	2	1	5
20	65	4	3	4	11	45	82	3	1	1	5
21	68	4	4	4	12	46	65	2	2	0	4
22	69	4	4	3	11	47	62	2	2	0	4
23	72	4	4	3	11	48	76	3	0	0	3
24	71	4	3	3	10	49	75	3	0	0	3
25	71	4	3	3	10	50	69	3	0	0	3
						51	68	3	0	0	3

2. Determine whether a significant relationship exists between class rank of the top 25 students at a public high school and hours per week involved in school-related sport activities by conducting a Kendall's tau-a and significance test using the formula in this unit. Report the findings by providing the two values. Use the data set provided.

Student	Class Rank	Hours Sports	Hours Ranked	C	D
1	1	6			
2	2	4			
3	3	0			
4	4	8			
5	5	12			
6	6	11			
7	7	9			
8	8	5			
9	9	7			
10	10	18			
				C =	D =

3. Determine whether a significant relationship exists between mathematics achievement score and annual household income rank/bracket by conducting a Spearman's rho test in SPSS using the results of a two-tailed test. Report the findings in an APA results section. Use the data set from exercise 5 in unit 3.

4. This exercise is based on data provided by Montgomery Co. of Maryland.

Following are the ages and size of dogs (by rank) still available to be adopted as of February 24, 2023, at the Montgomery County Animal Services & Adoption Center in Derwood, Maryland. Determine if a significant correlation exists between age and size with regard to dogs remaining to be adopted, with $\alpha = .05$. Use SPSS to derive the coefficient and p value. Explain your findings.

Size of dog rankings: 1 = small, 2 = medium, 3 = large, 4 = extra large

Age (in Months)	Size (Rank)
36	2
24	3
24	3
108	2
60	3
8	2
36	1
72	4
23	2
20	2

Age (in Months)	Size (Rank)
15	2
60	2
24	3
24	3
36	2
24	2
72	1
24	3
7	2
96	3
14	1
13	2
21	2
20	2
48	3
24	3
72	2

Analyzing and Reporting on Likert Scaling Data

OBJECTIVES

- Understand and apply the procedures to analyze Likert scaling data.
- Interpret the results and create an APA bar graph and results section with tables for Likert scaling data.

KEY TERMS

Median: The middle score of a set of scores when the scores are arranged in ascending order.

Bar chart: A chart using vertical bars to depict a distribution of values.

Receiver operator curve (ROC): A two-dimensional graph that plots a true positive rate against a false positive rate.

BRIEF INTRODUCTION TO THIS UNIT

In this unit, we will learn how to summate Likert scaling scores in SPSS, calculate the median score, calculate frequencies of scaling responses, and create a bar graph and tables for visual representation of the output. Oftentimes, a researcher wants an estimate of the mean score and 95 percent confidence intervals of the overall population on a construct. But unless we know the true underlying distribution, we cannot accurately calculate a population mean from a sample. This situation is made more difficult—if not impractical—when working with ordinal data for reasons mentioned in unit 1. Because Likert scaling data are ordinal, we will rely on the median score for estimating where subjects stand on an issue. These steps are just a few of many procedures and analyses that can be conducted in SPSS and similar software programs on ordinal data sets—most of which are *not* covered in this handbook. It is possible that your instructor will inform you of other analyses or demonstrate these other analyses as needed.[1]

For demonstration purposes, the assumption in this unit is that survey data were collected on the likelihood of randomly selected senior homeowners replacing grass lawns with xeriscaping. The data can be found in Unit 6, Exercise 1, and the population under study is senior homeowners living in the southwestern US. Let us also assume that, with data from previous administrations of this survey to senior homeowners *and* based on knowledge of which homeowners from prior surveys followed through and replaced grass lawns with xeriscaping, a demarcation[2] (cut-off) score of 14.00 was established. In other words, based on prior survey results and knowledge of which survey respondents followed up and installed xeriscaping, we can expect that homeowners who achieve a summated score of 14 or 15 are likely to follow through and replace grass lawns with xeriscaping.

ENTERING, SUMMATING, AND ANALYZING THE DATA IN SPSS

Let us look at the three variables (items) in the SPSS Variable View in table 7.1.

TABLE 7.1   Variable View in SPSS Data Field

	Name	Type	Width	Decimal	Label	Values	Missing	Columns	Align	Measure	Role
						Untitled 1 [Data Set0] – IBM SPSS Statistics Data Editor						
1	Item one	numerical	8	0	soon...	1 = very	none	8	right	ordinal	input
2	Item two	numerical	8	0	soon...	1 = very	none	8	right	ordinal	input
3	Item three	numerical	8	0	soon...	1 = very	none	8	right	ordinal	input

Source: Generated with SPSS software. Copyright © 2023 by IBM Corporation. Reprinted with permission.

---

1   An ordinal regression is often used when the researcher's dependent variable is ordinal and the researcher wishes to measure the strength of variables (e.g., gender, SES, age) to predict the ordinal outcome.
2   Demarcation (cut-off) scores are calculated by examining a receiver operator curve (ROC). A ROC can be easily conducted in SPSS if the researcher has both survey scores and knowledge/data on an outcome.

Be sure to enter the scaling in the Values box (see Appendix M) and set the Measure as ordinal. Now let us examine the scores of the first four subjects in the Data View field (table 7.2).

TABLE 7.2 Data View in SPSS Data Field

				Sum score
	item1	item2	item3	
1	5	5	5	15.00
2	5	5	5	15.00
3	5	5	5	15.00
4	5	5	4	15.00

Untitled 1 [Data Set0] – IBM SPSS Statistics Data Editor

Source: Generated with SPSS software. Copyright © 2023 by IBM Corporation. Reprinted with permission.

This is the point at which we want to command SPSS to summate subjects' scores.[3] SPSS will then create a new column for the summated scores, which we will refer to as the sumscore. Your instructor will demonstrate this procedure during a live lecture session.

Data Calculation and Analysis

The steps for commanding SPSS to calculate the median score and frequencies is as follows: ANALYZE > DESCRIPTIVE STATISTICS > FREQUENCIES. Move the scores of all five items as well as the *sumscore* variable into the Variables box. In Statistics, choose Median. Click Continue. In Charts, click Bar chart. Click Continue and click OK. Selected results can be found in table 7.3.

TABLE 7.3 Modified SPSS Output Items 1, 2, & 3 Output

		Frequency	Percent	Valid Percent	Cumulative Percent
Valid	very unlikely	0	0.0	0.0	0.0
	unlikely	3	5.9	5.9	5.9
	somewhat likely	20	39.2	39.2	45.1
	likely	14	27.5	27.5	72.5
	very likely	14	27.5	27.5	100.0
	Total	51	100.0	100.0	

Plan on soon taking steps

---

3 Summations can also be done in Excel before transferring the data set into SPSS.

Learn to appreciate the aesthetics				
		Frequency	Percent	Valid Percent	Cumulative Percent
Valid	very unlikely	14	27.5	27.5	27.5
	unlikely	11	21.6	21.6	49.0
	somewhat likely	20	39.2	39.2	88.2
	likely	0	0.0	0.0	88.2
	very likely	6	11.8	11.8	100.0
	Total	51	100.0	100.0	

Learn to enjoy working or spending time in a xeriscaping garden				
		Frequency	Percent	Valid Percent	Cumulative Percent
Valid	very unlikely	23	45.1	45.1	45.1
	unlikely	12	23.5	23.5	68.6
	somewhat likely	10	19.6	19.6	88.2
	likely	3	5.9	5.9	94.1
	very likely	3	5.9	5.9	100.0
	Total	51	100.0	100.0	

Source: Generated with SPSS software. Copyright © 2023 by IBM Corporation. Reprinted with permission.

In table 7.3, we can see the distribution of scores by percent for the 51 subjects on item 1: Soon take steps to reduce water consumption. Looking at the far-right column of Cumulative Percent, we can see that a majority of homeowners (54.9 percent) claim to be likely or very likely to begin taking steps to reduce water consumption. Attitudes begin to change for item 2: Learn to appreciate the aesthetics of xeriscaping. On this item, only 11.8 percent of homeowners claim to be likely or very likely to learn to appreciate the looks of xeriscaping. Attitudes are the same on item 3: Learn to enjoy working or spending time in a xeriscaping garden, when again only 11.8 percent of homeowners claim to be likely or very likely to enjoy their xeriscaping garden.

Since we know the cut-off score for seniors likely to act on replacement, it makes sense here to examine the percentages of sum scores. These can be found in table 7.4.

TABLE 7.4　Modified SPSS Output Frequency of Sums

	Sum Score			
	Frequency	Percent	Valid Percent	Cumulative Percent
Valid　4.00	3	5.9	5.9	5.9
5.00	11	21.6	21.6	27.5
6.00	7	13.7	13.7	41.2
7.00	3	5.9	5.9	47.1
8.00	2	3.9	3.9	51.0
9.00	9	17.6	17.6	68.6
10.00	2	3.9	3.9	72.5
11.00	8	15.7	15.7	88.2
14.00	3	5.9	5.9	94.1
15.00	3	5.9	5.9	100.0
Total	51	100.0	100.0	

Source: Generated with SPSS software. Copyright © 2023 by IBM Corporation. Reprinted with permission.

Upon examination of the cumulative percent column, we can expect 11.8 percent of these homeowners to replace their grass lawns with xeriscaping. So far, it appears that most seniors claim to have plans to replace grass lawns but with a small percentage of them likely to follow through on the claim. What does the median score tell us? For this sample population, the median sum score was 8.00,[4] in line with the other output and well below the cut-off score for actual replacement (14.00). Finally, an examination of a bar chart is called for, as it will provide a visualization of the sum scores (figure 7.1).

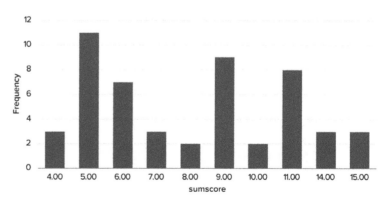

FIGURE 7.1.　Bar chart of summated scores.

Copyright © 2023 by International Business Machines Corporation (IBM). Reprinted with permission.

4　Not shown in this unit. Also not shown are the bar charts of the responses of the three individual items.

CREATING A RESULTS SECTION

Like bar charts, tables can facilitate the reader's understanding of the study's results. In the results section, we will provide the readers with a description of the sample population, results, and accompanying table:

> Fifty-one randomly chosen senior homeowners residing in the southwestern US were surveyed on the likelihood of replacing grass lawns with xeriscaping. The overall median score was 8.00,[5] from a range of three to 15. Results indicated that, while most subjects (54.9 percent) claim to plan on soon taking steps to reduce water consumption, only 11.8 percent were equal to or above the cut-off score of 14.00 for likelihood to follow through with xeriscaping. Frequencies for subjects likely or very likely to learn to appreciate xeriscaping and learn to enjoy working and spending time in xeriscaping were low (see table 7.5).

TABLE 7.5   Frequencies of Likert Scaling Responses to Xeriscaping in the Southwest US

Item	Scaling	Frequencies	%
Plan to soon take steps to reduce	Very unlikely	0	0
	Unlikely	3	5.9
	Somewhat likely	20	39.2
	Likely	14	27.5
	Very likely	14	27.5
Learn to appreciate xeriscaping	Very unlikely	14	27.5
	Unlikely	11	21.6
	Somewhat likely	20	39.2
	Likely	0	0.0
	Very likely	6	11.8
Enjoy working and spending time	Very unlikely	23	45.1
	Unlikely	12	23.5
	Somewhat likely	10	19.6
	Likely	3	5.9
	Very likely	3	5.9

CONCLUSION

Frequencies of responses and summated scores, the median score, tables, and bar charts are valuable measures for explaining Likert scaling responses on a construct. If available, cut-off scores also help the researcher and consumers of research to understand likelihoods of action or attitudes from survey results. These measures can all be calculated by using SPSS and other similar software programs.

---

5   Strictly for the purpose of comparison, the calculated mean score was 7.86.

Exercises

1. A researcher investigated the climate-change anxiety levels of 61 randomly selected South Florida high school students. Survey items were from the subset Personal Experience of the Climate Change Anxiety Scale (CCAS; Clayton and Karazsia 2020). Using SPSS, calculate the median summated score, frequencies of scaling responses for the three items, and frequencies of summated scores. Assume that psychologists consider a summated score of 14.00 or higher to be indicative of a student in need of counseling for their anxieties. Provide your answers in the form of a results section, including a table for the scaling responses for all three items. Use the data set provided.

1 = never, 2 = rarely, 3 = sometimes, 4 = often, 5 = almost always
Item 14: I have been directly affected by climate change.
Item 15: I know someone who has been directly affected by climate change.
Item 16: I have noticed a change in a place that is important to me due to climate change.

Subject	Item 14	Item 15	Item 16	Subject	Item 14	Item 15	Item 16
1	5	5	5	31	3	3	2
2	5	5	5	32	3	3	2
3	5	5	5	33	3	3	2
4	5	5	5	34	3	3	2
5	5	5	5	35	3	3	2
6	5	5	4	36	3	3	2
7	4	5	4	37	2	3	2
8	4	5	4	38	2	3	2
9	4	5	4	39	2	3	2
10	4	5	3	40	2	3	2
11	4	4	3	41	2	3	2
12	4	4	3	42	2	3	2
13	3	4	3	43	2	3	2
14	3	4	3	44	2	3	1
15	3	4	3	45	2	3	1
16	3	4	3	46	2	3	1
17	3	4	3	47	2	3	1
18	3	4	3	48	2	3	1
19	3	4	3	49	2	3	1

(Continued)

Subject	Item 14	Item 15	Item 16	Subject	Item 14	Item 15	Item 16
20	3	4	3	50	2	3	1
21	3	4	3	51	2	2	1
22	3	4	3	52	2	2	1
23	3	3	3	53	2	2	1
24	3	3	3	54	2	2	1
25	3	3	3	55	1	2	1
26	3	3	3	56	1	1	1
27	3	3	3	57	1	1	1
28	3	3	3	58	1	1	1
29	3	3	3	59	1	1	1
30	3	3	3	60	1	1	1
				61	1	1	1

2. Fifty instructional staff members from a large school district were surveyed to measure satisfaction levels of a newly proposed teacher evaluation system. Using SPSS, calculate the median summated score, frequencies of scaling responses for each of the four items, and frequencies of summated scores. Assume that a summated score of 17.00 or higher to be indicative of satisfaction of the newly proposed evaluation system. Provide your answers in the form of a results section, including a table for the scaling responses for all four items. Use the data set provided.

1 = very dissatisfied, 2 = dissatisfied, 3 = somewhat satisfied, 4 = satisfied, 5 = very satisfied
item1: How satisfied are you with the criteria for measuring teacher responsiveness?
item2: How satisfied are you with the criteria for measuring teacher question and discussion techniques?
item3: How satisfied are you with the criteria for measuring techniques to engage all students?

Subject	Item1	Item2	Item3	Subject	Item1	Item2	Item3
1	5	5	4	26	3	3	3
2	5	5	4	27	3	3	3
3	5	4	4	28	3	3	2
4	4	4	4	29	3	3	2
5	4	4	3	30	3	3	2
6	4	4	3	31	3	3	2
7	4	4	3	32	3	3	2
8	4	4	3	33	3	3	2

Subject	Item1	Item2	Item3	Subject	Item1	Item2	Item3
9	4	4	3	34	3	2	2
10	4	4	3	35	3	2	2
11	4	4	3	36	3	2	2
12	4	4	3	37	2	2	2
13	4	4	3	38	2	2	2
14	4	4	3	39	2	2	2
15	4	4	3	40	2	2	2
16	4	3	3	41	2	2	2
17	4	3	3	42	2	2	2
18	4	3	3	43	2	2	2
19	4	3	3	44	2	2	1
20	4	3	3	45	2	2	1
21	4	3	3	46	2	1	1
22	4	3	3	47	2	1	1
23	4	3	3	48	2	1	1
24	4	3	3	49	1	1	1
25	3	3	3	50	1	1	1

Answers to Unit Exercises

UNIT 1: INTRODUCTION TO ORDINAL DATA

1. The use of the descriptor *fair* might cause confusion (*lexical miscomprehension*) for the respondent and render this scaling nonordinal. In some dialects, the phrase *fair knowledge* can mean a good amount of knowledge. If the respondent fails to recognize the choice of fair knowledge as the second-least level of comprehension in the scaling, responses might be less reliable and this item a source of error.

2. When surveying students, possible answers include students having recently lost a family member or having recently undergone a traumatic event, students impaired by nonprescribed drugs or alcohol, second-language learners without translation services, or special needs students lacking the required services. Depending on the items and survey topic, parental consent may be required for students to complete the survey. When surveying patients, patients' physical and mental conditions must be optimal to the point of understanding the directions for reading the survey, completing the survey, and being able to ask questions about survey items if needed.

3. Answers may vary. Participation theory experts have identified several factors involved in response/nonresponse of general and/or academic surveys. The more salient the topic is to the individual, the more likely they will complete the survey. Institutional factors such as being a private institute versus a public school and a more positive peer effect (the social norm of the student population) can lead to higher response rates. Another factor is survey fatigue, in which students are saturated with survey requests to the point of not bothering to complete and submit the survey(s).

4. Some answers might be to (a) reduce the number of students asked to complete the evaluation via random sampling, (b) ask students to complete the evaluation only if the course is a part of their major studies (increasing the likelihood of salience), or (c) centralize the survey process at the institutional level to control survey fatigue.

5. Using Likert-like items can be a fast, efficient method of collecting data about sensory events. With the use of today's technology (e.g., email lists, class rosters, enrollment data), hundreds of potential subjects can be reached within minutes, reducing the need for face-to-face encounters, which can be difficult to arrange when the researcher wishes to collect hundreds or thousands of responses. But with online surveying comes many risks and limitations: the potential for response failure, the inability to ask follow-up questions, the inability to know exactly *who* completed the survey, and the inability to control for incomplete surveys.

UNIT 2: DISTRIBUTION-FREE ASPECTS OF NONPARAMETRIC TESTS

1. With $n = 10$ and $m = 10$, a W statistic of 121 ($x = 121$) produces a probability of .124. Fail to reject the null hypothesis. Scores are the same for both faculty and nonfaculty members.
2. With $n = 30$, a T^* value of 314 ($x = 314$) produces a probability of .0481. The manager can feel confident that the repairman improved his cleanliness when repairing appliances while on the job site.
3. With $k = 3$ (three groups), $n_1 = 8$, $n_2 = 8$, and $n_3 = 8$, and a calculated H value of 7.000, we lack evidence of any differences in pain scores between the three treatment groups. We would need an H value of 8.435 to reach a probability of .0101.

UNIT 3: THE RANK SUM TEST FOR TWO INDEPENDENT SAMPLES

1. If one set of scores were assigned the lowest ranks of 1–6, then $W_1 = 21$. If the other set of scores were assigned the higher, the remaining ranks of 7–12, then $W_2 = 57$. The difference would be 36.
2. When a small number of ties exist, the variance is reduced somewhat. This has little effect on the p value.
3. $W_{old} = 70$. $W_{new} = 35$. With $n = 7$ and $m = 7$ and $\alpha = .049$, $x (W_{cv}) = 66$.

   Reject the null hypothesis if $W_{new} \geq 66$.
   Reject the null hypothesis if $W_{new} \leq 39$.
   Since 35 is less than 39, we have evidence that the new model tire achieves less mileage.

4. $W_{GenZ} = 140$. $W_{BB} = 70$.
   Reject the null hypothesis if $z \geq 1.96$ or $z \leq -1.96$.

   $$z = 70 - 105 / 13.23$$
   $$z = -2.640$$

   Since –2.640 is to the left of –1.96, we reject the null hypothesis. We have evidence that baby boomers have significantly lower scores on climate change concerns.

5. A Wilcoxon rank sum test was conducted to evaluate the null hypothesis that no differences exist in mathematics achievement scores between two teaching–learning conditions. The test was significant, $z = -2.305$, $p = .021$. The mean rank for students under condition 2 (20.57) was significantly higher than the mean rank for students under condition 1 (12.91).

UNIT 4: THE SIGNED RANK TEST FOR PAIRED SAMPLES

1. When $n = 8$, $x\ (T_{cv}) = 30$ with $\alpha = .055$.

   Reject H_0: $\theta = 0$ in favor of H_1: $\theta \neq 0$ if $T^* \geq 30$ or $T^* \leq 8(9) / 2 - 30$.
   Reject the null hypothesis: $T^* = 4$ is less than 6.
   We have evidence that the new treatment leads to lower pain scores, $p \leq .055$.

2. First, adjust n to $n = 11$, as subject 6 showed no change.

   Reject H_0: $\theta = 0$ in favor of H_1: $\theta \neq 0$ if $z \geq 1.96$ or $z \leq 1.96$.

$$\mu T = 11(11 + 1) / 4 = 33$$
$$\sigma T = [11(11 + 1)(2 \times 11 + 1) / 24]^{.5} = 11.247$$
$$T^* = 59$$
$$z = 59 - 33 / 11.247$$
$$z = 2.311$$

Reject the null hypothesis, as 2.311 > than 1.96. Based on $z = 2.311$, $p = .021^*$ (see table A). Evidence exists that self-esteem scores increased as a result of attending the camp.

3. A Wilcoxon signed rank test was conducted to measure for significant change in cigarettes smoked from pre-hypnosis to post-hypnosis. The test was significant: $z = -2.106$, $p = .035$. The mean negative rank (5.50) was significantly greater than the positive mean rank (1.50). On average, subjects smoked fewer cigarettes posttreatment ($M = 24.22$, $SD = 9.46$) than pretreatment ($M = 28.22$, $SD = 6.76$) daily.

4. Any reduction in cigarette smoking should be welcomed. With that said, a daily four-cigarette reduction does not bring this population closer to quitting smoking. A population of casual smokers (one to five cigarettes daily) might have produced results that no cigarette was smoked daily posttreatment. Research shows that many factors play a role in quitting smoking, such as race, personality, peer and family pressure, and type of cigarette smoked. Researchers should describe the population under study to the degree that other researchers and professionals can make proper and accurate generalizations about the results of the study.

5. A Wilcoxon signed rank test was conducted to measure for significant change in nonviolent incidents among students under contract before and after contract modifications. The test was not significant, $z = -1.861$, $p = .063$.

UNIT 5: OMNIBUS TESTS FOR THREE OR MORE INDEPENDENT SAMPLES

1. When $k = 3$, all $n = 5$, and $\alpha = .0509$, $H_{cv} = 5.660$. Reject the null hypothesis if $H_{obs} \geq 5.660$.

   $\Sigma R_1 = 48$, $\Sigma R_2 = 23$, $\Sigma R_3 = 49$. $H_{obs} = 5.34$. Fail to reject the null hypothesis. We lack evidence of differences in the days to the presence of microorganisms between any of the pairs of levels of preservatives.

2. When $\alpha = .05$ and the degrees of freedom = 2, the X^2 critical values is 5.991 (see table E).

   $\Sigma R_1 = 93$, $\Sigma R_2 = 40$, $\Sigma R_3 = 38$; $\bar{R}_1 = 15.5$, $\bar{R}_2 = 6.7$, $\bar{R}_3 = 6.3$; $H_{obs} = 11.21$. Reject the null hypothesis for the omnibus test, as 11.21 is greater than the critical value of 5.991. Anxiety levels are significantly different with at least one pair of settings.

3. Fail to reject the null hypothesis. $H = 2.997$, $p = .223$. There is no evidence of differences in C-E-B scores between the three school programs.

UNIT 6: CORRELATIONAL MEASURES

1. Reject the null hypothesis. Kendall's $\tau = -.252$, $p = .013$.
2. Fail to reject the null hypothesis. Kendall's $\tau = -.333$, $z = 1.33$.

$$C - D / C + D = 30 - 15 / 30 + 15 = 15 / 45 = .333$$
$$z = .999\ (9.48) / 7.07 = 1.33$$

3. Fail to reject the null hypothesis. Spearman's $\rho = -.016$, $p = .931$.

UNIT 7: ANALYZING AND REPORTING ON LIKERT SCALING DATA

1. Sixty-one randomly chosen high school students residing in South Florida were surveyed on the frequency of personal experiences with climate change. The overall median score was 8.00, from a range of 3 to 15. Only 9.8 percent of high school students achieved summated scores indicative of needing counseling for anxieties. Frequencies for subjects on the three items can be found in the following table.

Item	Scaling	Frequencies	%
I have been directly affected	Never	7	11.5
	Rarely	18	29.5
	Sometimes	24	39.3
	Often	6	9.8
	Almost always	6	9.8
I know someone who has been affected	Never	6	9.8
	Rarely	5	8.2
	Sometimes	28	45.9
	Often	12	19.7
	Almost always	10	16.4
I have noticed a change in a place	Never	18	29.5
	Rarely	13	21.3
	Sometimes	21	34.4
	Often	4	6.6
	Almost always	5	8.2

2. Fifty randomly chosen teachers from a large school district in the Midwest US were surveyed on their satisfaction for the criteria from a portion of a newly proposed teacher evaluation system. The overall median score was 9.00, short of the cut-off score of 11.00 for likelihood of voting for the proposed system. Only 30 percent of the teachers surveyed demonstrated levels of satisfaction $\geq$ 11.00. Frequencies for teachers on the three items can be found in the following table.

Item	Scaling	Frequencies	%
Measuring teacher responsiveness	Very dissatisfied	2	4
	Dissatisfied	12	24
	Somewhat satisfied	12	24
	Satisfied	21	42
	Very satisfied	3	6
Measuring question/discussion	Very dissatisfied	5	10
	Dissatisfied	12	24
	Somewhat satisfied	18	36
	Satisfied	13	26
	Very satisfied	2	4
Measuring engagement of students	Very dissatisfied	7	14
	Dissatisfied	16	32
	Somewhat satisfied	23	46
	Satisfied	4	8
	Very satisfied	0	0

Tables

TABLE A Upper-Tail Probabilities of the z-Distribution

z	.00	.01	.02	.03	.04	.05	.06	.07	08.	.09
0.0	.5000	.4960	.4920	.4880	.4840	.4801	.4761	.4721	.4681	.4641
0.1	.4602	.4562	.4522	.4483	.4443	.4404	.4364	.4325	.4286	.4247
0.2	.4207	.4168	.4129	.4090	.4052	.4013	.3974	.3936	.3897	.3859
0.3	.3821	.3783	.3745	.3707	.3669	.3632	.3594	.3557	.3520	.3483
0.4	.3446	.3409	.3372	.3336	.3300	.3264	.3228	.3192	.3156	.3121
0.5	.3085	.3050	.3015	.2981	.2946	.2912	.2877	.2843	.2810	.2776
0.6	.2743	.2709	.2676	.2643	.2611	.2578	.2546	.2514	.2483	.2451
0.7	.2420	.2389	.2358	.2327	.2296	.2266	.2236	.2206	.2177	.2148
0.8	.2119	.2090	.2061	.2033	.2005	.1977	.1949	.1922	.1894	.1867
0.9	.1841	.1814	.1788	.1762	.1736	.1711	.1685	.1660	.1635	.1611
1.0	.1587	.1562	.1539	.1515	.1492	.1469	.1446	.1423	.1401	.1379
1.1	.1357	.1335	.1314	.1292	.1271	.1251	.1230	.1210	.1190	.1170
1.2	.1151	.1131	.1112	.1093	.1075	.1056	.1038	.1020	.1003	.0985
1.3	.0968	.0951	.0934	.0918	.0901	.0885	.0869	.0853	.0838	.0823
1.4	.0808	.0793	.0778	.0764	.0749	.0735	.0721	.0708	.0694	.0681
1.5	.0668	.0655	.0643	.0630	.0618	.0606	.0594	.0582	.0571	.0559
1.6	.0548	.0537	.0526	.0516	.0505	.0495	.0485	.0475	.0465	.0455
1.7	.0446	.0436	.0427	.0418	.0409	.0401	.0392	.0384	.0375	.0367
1.8	.0359	.0351	.0344	.0336	.0329	.0322	.0314	.0307	.0301	.0294
1.9	.0287	.0281	.0274	.0268	.0262	.0256	.0250	.0244	.0239	.0233
2.0	.0228	.0222	.0217	.0212	.0207	.0202	.0197	.0192	.0188	.0183

(*Continued*)

z	.00	.01	.02	.03	.04	.05	.06	.07	08.	.09
2.1	.0179	.0174	.0170	.0166	.0162	.0158	.0154	.0150	.0146	.0143
2.2	.0139	.0136	.0132	.0129	.0125	.0122	.0119	.0116	.0113	.0110
2.3	.0107	.0104	.0102	.0099	.0096	.0094	.0091	.0089	.0087	.0084
2.4	.0082	.0080	.0078	.0075	.0073	.0071	.0069	.0068	.0066	.0064
2.5	.0062	.0060	.0059	.0057	.0055	.0054	.0052	.0051	.0049	.0048
2.6	.0047	.0045	.0044	.0043	.0041	.0040	.0039	.0038	.0037	.0036
2.7	.0035	.0034	.0033	.0032	.0031	.0030	.0029	.0028	.0027	.0026
2.8	.0026	.0025	.0024	.0023	.0023	.0022	.0021	.0021	.0020	.0019
2.9	.0019	.0018	.0018	.0017	.0016	.0016	.0015	.0015	.0014	.0014
3.0	.0013	.0013	.0013	.0012	.0012	.0011	.0011	.0011	.0010	.0010
3.1	.0010	.0009	.0009	.0009	.0008	.0008	.0008	.0008	.0007	.0007
3.2	.0007	.0007	.0006	.0006	.0006	.0006	.0006	.0005	.0005	.0005
3.3	.0005	.0005	.0005	.0004	.0004	.0004	.0004	.0004	.0004	.0003
3.4	.0003	.0003	.0003	.0003	.0003	.0003	.0003	.0003	.0003	.0002

Source: Dennis E. Hinkle, William Wiersma and Stephen G. Jurs, "Standard Normal Curve of z Values," *Applied Statistics for the Behavioral Sciences*, p. 633. Copyright © 1998 by HarperCollins Publishers.

TABLE B   Selected Upper-Tail Probabilities for the Null Distribution of the Wilcoxon Rank Sum *W* Statistic

	n = 5					
x	m = 5	m = 6	m = 7	m = 8	m = 9	m = 10
28	.500					
29	.421					
30	.345	.353				
31	.274	.465				
32	.210	.396				
33	.155	.331	.500			
34	.111	.268	.438			
35	.075	.214	.378	.528		
36	.048	.165	.319	.472		
37	.028	.123	.265	.416		
38	.016	.089	.216	.362	.500	
39	.008	.063	.172	.311	.449	

			$n = 5$			
x	m = 5	m = 6	m = 7	m = 8	m = 9	m = 10
40	.004	.041	.134	.262	.399	.523
41		.026	.101	.218	.350	.477
42		.015	.074	.177	.303	.430
43		.009	.053	.142	.259	.384
44		.004	.037	.111	.219	.339
45		.002	.024	.085	.182	.297
46			.015	.064	.149	.257
47			.009	.047	.120	.220
48			.005	.033	.095	.185
49			.003	.023	.073	.155
50			.001	.015	.056	.127
51				.009	.041	.103
52				.005	.030	.082
53				.003	.021	.065
54				.002	.014	.050
55				.001	.009	.038
56					.006	.028
57					.003	.020
58					.002	.014
59					.001	.010
60					.000	.006
61						.004
62						.002
63						.001
64						.001
65						.000

			$n = 6$		
x	m = 6	m = 7	m = 8	m = 9	m = 10
39	.531				
40	.469				
41	.409				
42	.350	.527			
43	.294	.473			
44	.242	.418			
45	.197	.365	.525		

(Continued)

			$n = 6$		
x	m = 6	m = 7	m = 8	m = 9	m = 10
46	.155	.314	.475		
47	.120	.267	.426		
48	.090	.223	.377		
49	.066	.183	.331		
50	.047	.147	.286	.432	
51	.032	.117	.245	.388	.521
52	.021	.090	.207	.344	.479
53	.013	.069	.172	.303	.437
54	.008	.051	.141	.264	.396
55	.004	.037	.114	.228	.356
56	.002	.026	.091	.194	.318
57	.001	.017	.071	.164	.281
58		.011	.054	.136	.246
59		.007	.041	.112	.214
60		.004	.030	.091	.184
61		.002	.021	.072	.157
62		.001	.015	.057	.132
63		.001	.010	.044	.110
64			.006	.033	.090
65			.004	.025	.074
66			.002	.018	.059
67			.001	.013	.047
68			.001	.009	.036
69			.000	.006	.028
70				.004	.021
71				.002	.016
72				.001	.011
73				.001	.008
74				.000	.005
75				.000	.004
76					.002
77					.001
78					.001
79					.000

	n = 7					n = 8		
x	m = 7	m = 8	m = 9		x	m = 8	m = 9	m = 10
53	.500				68	.520		
54	.451				69	.480		
55	.402				70	.439		
56	.355	.522			71	.399		
57	.310	.478			72	.360	.519	
58	.267	.433			73	.323	.481	
59	.228	.389			74	.287	.444	
60	.191	.347	.500		75	.253	.407	
61	.159	.306	.459		76	.221	.371	.517
62	.130	.268	.419		77	.191	.336	.483
63	.104	.232	.379		78	.164	.303	.448
64	.082	.198	.340		79	.139	.271	.414
65	.064	.168	.303		80	.117	.240	.381
66	.049	.140	.268		81	.097	.212	.348
67	.036	.116	.235		82	.080	.185	.317
68	.027	.095	.204		83	.065	.161	.286
69	.019	.076	.176		84	.052	.138	.257
70	.013	.060	.150		85	.041	.118	.230
71	.009	.047	.126		86	.032	.100	.204
72	.006	.036	.105		87	.025	.084	.180
73	.003	.027	.087		88	.019	.069	.158
74	.002	.020	.071		89	.014	.057	.137
75	.001	.014	.057		90	.010	.046	.118
76	.001	.010	.045		91	.007	.037	.102
77	.000	.007	.036		92	.005	.030	.086
78		.005	.027		93	.003	.023	.073
79		.003	.021		94	.002	.018	.061
80		.002	.016		95	.001	.014	.051
81		.001	.011		96	.001	.010	.042
82		.001	.008		97	.001	.008	.034
83		.000	.006		98	.000	.006	.027
84		.000	.004		99	.000	.004	.022
85			.003		100	.000	.003	.017
86			.002		101		.002	.013

(*Continued*)

n = 7			
x	m = 7	m = 8	m = 9
87			.001
88			.001
89			.000
90			.000
91			.000

n = 8			
x	m = 8	m = 9	m = 10
102		.001	.010
103		.001	.008
104		.000	.006
105		.000	.004
106		.000	.003
107		.000	.002
108		.000	.002
109			.001

n = 9		
x	m = 9	m = 10
90	.365	.516
91	.333	.484
92	.302	.452
93	.273	.421
94	.245	.390
95	.218	.360
96	.193	.330
97	.170	.302
98	.149	.274
99	.129	.248
100	.111	.223
101	.095	.200
102	.081	.178
103	.068	.158
104	.057	.139
105	.047	.121
106	.039	.106
107	.031	.091
108	.025	.078
109	.020	.067
110	.016	.056

n = 10	
x	m = 10
105	.515
106	.485
107	.456
108	.427
109	.398
110	.370
111	.342
112	.315
113	.289
114	.264
115	.241
116	.218
117	.197
118	.176
119	.157
120	.140
121	.124
122	.109
123	.095
124	.083
125	.072

	n = 9				n = 10	
x	m = 9	m = 10			x	m = 10
111	.012	.047			126	.062
112	.009	.039			127	.053
113	.007	.033			128	.045
114	.005	.027			129	.038
115	.004	.022			130	.032
116	.003	.017			131	.026
117	.002	.014			132	.022
118	.001	.011			133	.018
119	.001	.009			134	.014
120	.001	.007			135	.012
121	.000	.005			136	.009
122	.000	.004			137	.007
123	.000	.003			138	.006
124	.000	.002			139	.004
125	.000	.001			140	.003
126	.000	.001			141	.003
127		.001			142	.002
128		.000			143	.001
129		.000			144	.001
130		.000			145	.001
131		.000			146	.001

Source: Myles Hollander and Douglas A. Wolfe, "Wilcoxon Rank Sum W Statistic," *Nonparametric Statistical Methods,* pp. 589, 590. Copyright © 1999 by John Wiley & Sons, Inc.m

TABLE C Selected Upper-Tail Probabilities for the Null Distribution of the Wilcoxon Signed Rank T^+ Statistic

	x	p			x	p
n = 3	3	.625		n = 7	21	.148
	4	.375			22	.109
	5	.250			23	.078
	6	.125			24	.055
					25	.039
n = 4	5	.562			26	.023
	6	.438			27	.016
	7	.312			28	.008
	8	.188				
	9	.125		n = 8	18	.527

(*Continued*)

TABLE C *(Continued)*

	x	p
	10	.062
$n = 5$	8	.500
	9	.406
	10	.312
	11	.219
	12	.156
	13	.094
	14	.062
	15	.031
$n = 6$	11	.500
	12	.422
	13	.344
	14	.281
	15	.219
	16	.156
	17	.109
	18	.078
	19	.047
	20	.031
	21	.016
$n = 7$	14	.531
	15	.469
	16	.406
	17	.344
	18	.289
	19	.234
	20	.188
$n = 9$	34	.102
	35	.082
	36	.064
	37	.049
	38	.037
	39	.027

	x	p
	19	.473
	20	.422
	21	.371
	22	.320
	23	.273
	24	.230
	25	.191
	26	.156
	27	.125
	28	.098
	29	.074
	30	.055
	31	.039
	32	.027
	33	.020
	34	.012
	35	.008
	36	.004
$n = 9$	23	.500
	24	.455
	25	.410
	26	.367
	27	.326
	28	.285
	29	.248
	30	.213
	31	.180
	32	.150
	33	.125
$n = 10$	40	.116
	41	.097
	42	.080
	43	.065
	44	.053
	45	.042

	x	p			x	p
	40	.020			46	.032
	41	.014			47	.024
	42	.010			48	.019
	43	.006			49	.014
	44	.004			50	.010
	45	.002			51	.007
					52	.005
n = 10	28	.500			53	.003
	29	.461			54	.002
	30	.423			55	.001
	31	.385				
	32	.348				
	33	.312				
	34	.278				
	35	.246				
	36	.216				
	37	.188				
	38	.161				
	39	.138				
	40	.116				
	41	.097				
	42	.080				
	43	.065				
	44	.053				
	45	.042				
	46	.032				
	47	.024				
	48	.019				
	49	.014				
	50	.010				
	51	.007				
	52	.005				
	53	.003				
	54	.002				
	55	.001				

Source: Myles Hollander and Douglas A. Wolfe, "Wilcoxon Signed Rank T Statistic," *Nonparametric Statistical Methods*, pp. 576, 577, 579. Copyright © 1999 by John Wiley & Sons, Inc.

TABLE D Selected Upper-Tail Probabilities for the Null Distribution of the Kruskal–Wallis H Statistic

n_1	n_2	n_3	x	p	n_1	n_2	n_3	x	p
3	3	3	4.622	.1000	3	3	6	5.038	.0748
			5.067	.0857				5.551	.0512
			5.422	.0714				5.615	.0497
			5.600	.0500				6.385	.0253
			5.956	.0250				6.436	.0223
			6.489	.0107				7.192	.0102
			7.200	.0036				7.410	.0078
								7.615	.0061
3	3	4	4.700	.1010				7.872	.0043
			4.709	.0924				8.692	.0010
			4.845	.0810					
			5.000	.0743	3	4	4	4.477	.1022
			5.727	.0505				4.545	.0091
			5.791	.0457				5.053	.0781
			6.018	.0267				5.144	.0729
			6.155	.0248				5.576	.0507
			6.745	.0100				5.598	.0487
			7.000	.0062				6.386	.0262
			7.318	.0043				6.394	.0248
			8.018	.0014				7.136	.0107
								7.144	.0097
3	3	5	4.412	.1091				7.477	.0062
			4.533	.0970				7.898	.0042
			5.042	.0775				8.326	.0012
			5.079	.0693				8.909	.0005
			5.515	.0507					
			5.648	.0489	3	4	5	4.523	.1033
			6.303	.0255				4.549	.0989
			6.315	.0212				4.538	.0754
			6.982	.0113				4.953	.0742
			7.079	.0087				5.631	.0504
			7.515	.0054				5.656	.0486
			7.636	.0041				6.410	.0250

					$k = 3$					
n_1	n_2	n_3	x	p		n_1	n_2	n_3	x	p
			8.242	.0011					7.395	.0109
			8.727	.0007					7.445	.0097
									7.927	.0050
3	3	6	4.538	.1034					8.626	.0012
			4.590	.0977					8.795	.0009
			4.949	.0770						
3	4	6	4.604	.1000		5	5	5	4.940	.0807
			4.962	.0769					5.040	.0746
			5.033	.0744					5.660	.0509
			5.604	.0504					5.780	.0488
			5.610	.0486					6.720	.0259
			6.538	.0250					6.740	.0248
			7.467	.0101					7.980	.0105
			7.500	.0097					8.000	.0095
			8.033	.0050					8.780	.0050
			9.170	.0010					9.920	.0010
3	5	5	4.536	.1020		6	6	6	4.538	.1010
			4.545	.0997					4.643	.0987
			4.993	.0755					5.064	.0759
			5.020	.0720					5.099	.0742
			5.626	.0508					5.719	.0502
			5.705	.0461					5.801	.0491
			6.488	.0254					6.877	.0259
			6.549	.0244					6.889	.0249
			7.543	.0102					8.187	.0102
			7.578	.0097					8.222	.0099
			8.264	.0051					9.088	.0050
			8.316	.0049					10.819	.0010
			9.284	.0010						
						7	7	7	4.549	.1007
4	4	4	4.500	.1042					4.594	.0993
			4.654	.0966					5.076	.0750
			4.962	.0800					5.766	.0506
			5.115	.0741					5.819	.0491
			5.654	.0546					6.909	.0256

(*Continued*)

n_1	**n_2**	**n_3**	**x**	**p**	**n_1**	**n_2**	**n_3**	**x**	**p**
			5.692	.0487				6.954	.0245
			6.577	.0263				8.334	.0101
			6.615	.0242				8.378	.0099
			7.538	.0107				9.358	.0051
			7.654	.0076				9.373	.0049
			7.731	.0066				11.288	.0010
			8.000	.0046					
			8.769	.0012					
			9.269	.0005					
5	5	5	4.500	.1015					
			4.560	.0995					

Source: Myles Hollander and Douglas A. Wolfe, "Kruskal Wallis H Statistic," *Nonparametric Statistical Methods*, pp. 641, 642. Copyright © 1999 by John Wiley & Sons, Inc.

TABLE E Selected Upper Percentage Points of the χ^2 Distribution

df	.10	.05	.02	.01	.001
1	2.706	3.841	5.412	6.635	10.827
2	4.605	5.991	7.824	9.210	13.815
3	6.251	7.815	9.837	11.325	16.266
4	7.779	9.488	11.668	13.277	18.467
5	9.236	11.070	13.388	15.086	20.515
6	10.645	12.592	15.033	16.812	22.457
7	12.017	14.067	16.622	18.475	24.322
8	13.362	15.507	18.168	20.090	26.125
9	14.684	16.919	19.679	21.666	27.877
10	15.987	18.307	21.161	23.209	29.588
11	17.275	19.675	22.618	24.725	31.264
12	18.549	21.026	24.054	26.217	32.909
13	19.812	22.362	25.472	27.688	34.528
14	21.064	23.685	26.873	29.141	36.123
15	22.307	24.996	28.259	30.578	37.697
16	23.542	26.296	29.633	32.000	39.252
17	24.769	27.587	30.995	33.409	40.790

df	.10	.05	.02	.01	.001
18	25.989	28.869	32.346	34.805	42.312
19	27.204	30.144	33.687	36.191	43.820
20	28.412	31.410	35.020	37.566	45.315
21	29.615	32.671	36.343	38.932	46.797
22	30.813	33.924	37.659	40.289	48.268
23	32.007	35.172	38.968	41.638	49.728
24	33.196	36.415	40.270	42.980	51.179
25	34.382	37.652	41.566	44.314	52.620
26	35.563	38.885	42.856	45.642	54.052
27	36.741	40.113	44.140	46.963	55.476
28	37.916	41.337	45.419	48.278	56.893
29	39.087	42.557	46.693	49.558	58.302
30	40.256	43.773	47.962	50.892	59.703

Source: Dennis E. Hinkle, William Wiersma and Stephen G. Jurs, *Applied Statistics for the Behavioral Sciences*, p. 638.
Copyright © 1998 by HarperCollins Publishers.

Bibliography

American Psychological Association. 2020. *Publication manual of the American Psychological Association.* Washington, DC: Author.

Clayton, S., and B. Karazsia. 2020. "Development and Validation of a Measure of Climate Change Anxiety." *Journal of Environmental Psychology 69* (2020): 101434. https://doi.org10.1016/j.jenvp.2020.101434

Daily Racing Form. 2022. Equibase.

Harter, H. L., and D. B. Owen, eds. 1973. *Selected Tables in Mathematical Statistics* (vol. 1). Chicago: Markham.

Harter, H. L., and D. B. Owen, eds. 1975. *Selected Tables in Mathematical Statistics* (vol. 3). Chicago: Markham.

Hinkle, D. E., W. Wiersma, and S. G. Jurs. 1998. *Applied Statistics for the Behavioral Sciences.* 4th ed. Boston: Houghton Mifflin.

Hollander, M., and D. A. Wolfe. 1999. *Nonparametric Statistical Methods.* 2nd ed. New York: Wiley-Interscience.

Iman, R. L., D. Quade, and D. A. Alexander. "Exact Probability Levels for the Kruskal-Wallis Test." In *Selected Tables in Mathematics Statistics* (vol. 3). Edited by H. L. Harter and D. B. Owen, 329–384. Chicago: Markham, 1975.

International Business Machines. 2019. *Statistical Package for the Social Sciences, version 27.* Author.

Kraft, C. H., and C. van Eeden. 1968. *A Nonparametric Introduction to Statistics.* New York: Macmillan.

Marcantonio, R., and J. Peck. 2022. "Correspondences from Quality Assurance." IBM.

Payen, J. F., O. Bru, J. L. Bosson, A. Lagrasta, E. Novel, I. Deschaux, P. Lavagne, & C. Jacquot, C. (2002). "Assessing Pain in Critically Ill Sedated Patients by Using a Behavioral Pain Scale." *Critical Care Medicine* 29, no. 12 (December 2001): 2258–63.

Ramsey, P. H. 1989. Critical values for Spearman's rank order correlation. *Journal of Educational Statistics 14,* no. 3 (1989): 245–53.

van Sonderen, Eric, Robbert Sanderman, and James C. Coyne. 2013. "Ineffectiveness of Reverse Wording of Questionnaire Items: Let's Learn from Cows in the Rain." *PLoS ONE* 8, no. 7 (July 2013): 1–6. https://journals.plos.org/plosone/article?id=10.1371/journal.pone.0068967

Wilcoxon, F., S. K. Katti, and R. A. Wilcox. 1973. "Critical Values and Probability Levels for the Wilcoxon Rank Sum Test and the Wilcoxon Signed Rank Test." In *Selected Tables in Mathematical Statistics* (vol. 1), edited by H. L. Harter and D. B. Owen, 171–260. Chicago: Markham.